PETER WADHAMS

A Farewell To Ice

with a Foreword by Walter Munk

ALLEN LANE
an imprint of
PENGUIN BOOKS

ALLEN LANE

UK | USA | Canada | Ireland | Australia
India | New Zealand | South Africa

Allen Lane is part of the Penguin Random House group of companies
whose addresses can be found at global.penguinrandomhouse.com

First published 2016
001

Copyright © Peter Wadhams, 2016

The moral right of the author has been asserted

Set in 10.5/14 pt Sabon LT Std
Typeset by Jouve (UK), Milton Keynes
Printed in Great Britain by Clays Ltd, St Ives plc

A CIP catalogue record for this book is available from the British Library

ISBN: 978-0-241-00941-3

In memory of old Arctic friends

Bill Campbell
Max Coon
Norman Davis
Moira and Max Dunbar
Geoff Hattersley-Smith
Wally Herbert
Lyn Lewis
Ray Lowry
Nobuo Ono
Erkki Palosuo
Gordon Robin
Unsteinn Stefánsson
Charles Swithinbank
Norbert Untersteiner
Thomas Viehoff

Contents

List of Plates

List of Figures

Every effort has been made to contact all copyright holders. The publishers will be happy to make good in future editions any errors or omissions brought to their attention.

Acknowledgements

I am grateful to innumerable people who have helped with facts, ideas and inspirations. They include Paul Beckwith, Peter Carter, Florence Fetterer, Martin Harrison, Chris Hope, Charles Kennel, Daniel Kieve, Seelye Martin, Walter Munk, Jon Nissen, Jim Overland, Hans Joachim Schellnhuber, David Wasdell and Gail Whiteman. I am grateful to Carl Wunsch, David Wasdell and Subhankar Banerjee for reading the whole manuscript and suggesting valuable additions and changes. I thank the US Office of Naval Research for long-term scientific support which made this book possible; also Andrea Pizzuti of Grafiche Fioroni, Casette d'Ete (Fermo), Italy, for help with illustrations. Most of all I thank my wife, Maria Pia Casarini, Director of the Istituto Geografico Polare 'Silvio Zavatti', Fermo, Italy, for a lifetime of personal support and inspiration.

I have named this book *A Farewell to Ice* (with apologies to Ernest Hemingway) because it deals not only with the implications for our planet of the huge loss of sea ice which is currently occurring, but also interleaves it with some personal stories from a long career which I hope shed light on the special nature of the world of sea ice and the consequences of its disappearance.

Chapter 12 draws on a review originally published as 'Antarctic Sea Ice Changes and their Implications: The Annual Ice Cycle and its Changes', in *Climate Change: Observed Impacts on Planet Earth*, 2nd edition, ed. Trevor Letcher (Amsterdam: Elsevier, 2015).

Foreword

Peter Wadhams has been a polar researcher for forty-seven years, during which time he has observed and measured major changes in the nature of sea ice cover in the polar regions. In this book, he first gives a brief review of the planet and the development of its ice on land and sea. He goes on to describe the profound transformations that he has witnessed during his career. The area of Arctic sea ice in summer has dwindled from more than 8 million square kilometres to less than half that, leading to projections for the imminent occurrence of ice-free summers.

The melting of sea ice is not just a curious phenomenon in a remote part of our world: it sharply decreases the amount of solar radiation reflected back into space, from 60 per cent to 10, further accelerating the planetary warming cycle. Frozen sediments, which have lain undisturbed since the last Ice Age, are now releasing plumes of methane – a very potent greenhouse gas – into the atmosphere. *A Farewell to Ice* is both an authoritative report on the state of the Arctic today and a timely reminder of the global threat posed by the loss of its sea ice.

Walter Munk
Scripps Institution of Oceanography, La Jolla, California

I

Introduction: A Blue Arctic

I have been a polar researcher since 1970. For most of those years I had the privilege of being based at the Scott Polar Research Institute in Cambridge, and served as its Director. Established as a memorial to Captain Robert Falcon Scott, it was a haven and meeting place for polar researchers of every kind of discipline, many of whom took long leaves of absence from their host institutions to study in its incomparable library. Throughout the 1970s and 1980s I was in the polar regions (usually the Arctic) every year, sometimes several times, and, like my colleagues in Europe, America, Russia and Japan, devoted myself to understanding the basic physical processes which occur in sea ice and which determine how it grows, decays and moves. Field research into ice is very difficult, and at times dangerous, and few of us considered that the object of our efforts, the Arctic Ocean, could be changing before our eyes. It was hard enough trying to understand how the Arctic worked in the first place. Yet changing it was. I was fortunate enough to be one of the first to obtain definite evidence of this, when I compared surveys of ice thickness that I had made from submarines in 1976 and 1987 and found a 15 per cent loss of average thickness. This result, published in *Nature* in 1990,[1] stimulated more intensive work which, within a decade, showed that this thinning was not only real, but that the ice had by then thinned by more than 40 per cent since the 1970s.[2] Something truly dramatic was going on. Polar researchers raised their eyes from their own specialized studies and started to consider the larger picture. They have become climate change specialists, indeed climate change pioneers, since it is in the Arctic that global change appears to be most rapid and drastic.

My interest in the polar oceans had been stimulated when I first went to the Arctic in the summer of 1970, aboard the Canadian oceanographic ship *Hudson*, which was carrying out the first circumnavigation of the Americas. The Hudson-70 expedition had left Nova Scotia in the chilly autumn of 1969 and had already sailed down to the Antarctic Peninsula, the Southern Ocean, the fjords of Chile and the vast expanse of the Pacific.[3] Now we were to attempt a feat accomplished by only nine ships before us, a transit of the Northwest Passage.[4] The ship was ice strengthened, and needed to be. All along the north coasts of Alaska and the Northwest Territories, the Arctic Ocean sea ice lay close in to the land, leaving us a channel of open water only a few miles wide to carry out our surveys. Sometimes the ice pushed right up to the coast and we had to break our way through densely clustered floes of heavy, thick multi-year ice (Plate 1), and eventually, when we were in the middle of the Northwest Passage, we had to be rescued by a heavy Government icebreaker, the *John A. Macdonald*. In those days, a battle with sea ice in the Canadian Arctic was considered normal. In 1903–6 it took Amundsen three years to navigate the Northwest Passage, and the second ship to make the Passage, the Royal Canadian Mounted Police schooner *St. Roch*, needed two seasons in 1942–4.

Today a ship entering the Arctic from Bering Strait in summer finds an ocean of open water in front of her. This blue water extends far to the north, stopping not far short of the North Pole. By the time this book is published it is possible – and, according to many predictions, likely – that the Pole itself will be uncovered for the first time in tens of thousands of years. The Northwest Passage is now easily navigable, and by the end of 2015 a total of 238 ships had sailed through it. In September 2012 sea ice covered only 3.4 million square kilometres (km^2) of the Arctic Ocean's surface, down from 8 million km^2 in the 1970s. It is difficult to overstate what this means in terms of planetary change. Our planet has actually changed colour. We all remember the first beautiful photograph of planet Earth rising from behind the Moon, taken by the Apollo-8 astronauts, a delicate blue sphere, isolated in the cosmos, which contains all that we know of life. That sphere was white at both ends. Today, from space, the top of the world in the northern summer looks blue instead of white. We have created an ocean where there was once an ice sheet. It is Man's first major achievement in reshaping the face of his planet, and it is of course an unintended achievement, with dubious and possibly catastrophic consequences to follow.

Things are actually even worse than appearances would suggest. My sonar measurements showed that the average ice thickness dropped 43 per cent between 1976 and 1999.[5] But they showed other things as well. In the past most of the ice in the Arctic was several years old, what is called *multi-year ice*. It had a rugged and magnificent topography, with huge pressure ridges which blocked the paths of explorers and which had keels reaching down 50 metres or more into the ocean (fig. 1.1). During the last decade a changing current system has driven most of this ice out of the Arctic, and it has been replaced by first-year ice (Plate 4), which grows during a single winter season, reaching a maximum thickness of only 1.5 metres, and has only a few shallow ridges to break up the very flat ice surface. Ice which grows to this lesser thickness during a single winter can melt away completely during a single summer because of warmer air and sea temperatures. It will not be long before the summer melt outstrips the winter growth everywhere in the Arctic, and when that happens the entire remaining summer ice cover will collapse. We will have

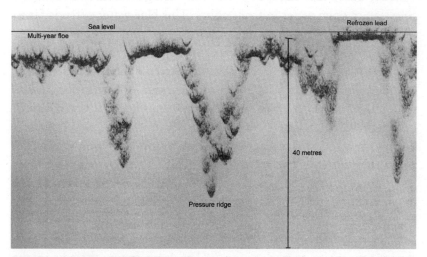

Figure 1.1: Pressure ridges in multi-year ice, recorded by upward-looking sonar from a submarine. The largest is 30 metres deep.

entered what the US climatologist Mark Serreze called the 'Arctic death spiral'.[6] In the very near future, as I explain in Chapter 7, it will leave us with an ice-free September in the Arctic, and within a few years after that the ice-free season will expand to four or five months.

The consequences of a collapse of Arctic summer ice will be dramatic. Two huge effects will be unleashed. First, once summer ice yields to open water, the *albedo* – the fraction of incoming solar radiation which is reflected straight back into space – drops from 0.6 to 0.1, which will further accelerate warming of the Arctic and of the whole planet. The albedo change from the loss of the last 4 million km² of ice will have the same warming effect on the Earth as the last twenty-five years of carbon dioxide emissions. Secondly, removal of the ice cover will take away a vital air-conditioning system for the Arctic. So long as some ice is present in summer, however thin, the sea surface temperature cannot rise above 0°C, since any warmer water loses its heat in melting some of the overlying ice. When the overlying ice is gone, the surface water can warm up by several degrees in summer (satellite observations have shown 7°C), and over the shallow continental shelves wind-induced mixing extends this heat down to the seabed. This then thaws the surface layer of the offshore permafrost, frozen seabed sediments which have lain there undisturbed since the last Ice Age. The thawing offshore permafrost will trigger the release of huge plumes of methane from the disintegration of methane hydrates trapped in the sediment. Methane has a greenhouse warming effect twenty-three times greater per molecule than carbon dioxide. An annual Russian–US expedition to the East Siberian Sea has already observed methane plumes welling up from the seabed, while other expeditions have seen methane plumes in the Laptev and Kara seas. If this release causes general atmospheric levels of the gas to rise, it will give a further immediate boost to global warming. I have written this book to explain these dramatic changes, and how and why the loss of Arctic ice is a threat to us all, not just an interesting change happening in a remote part of the world.

I have spent my entire scientific life, from the age of twenty-one, working on the science of sea ice and the polar oceans. What do these changes mean to me as I prepare to say a personal farewell to this magical landscape? Overwhelmingly I feel that this is a spiritual impoverishment of the Earth as well as a practical catastrophe for mankind. Our own greed and stupidity are taking away the beautiful world of Arctic Ocean sea ice, which once protected us from the impacts of climatic extremes. Now urgent action is needed if we are to save ourselves from the consequences.

2

Ice, the Magic Crystal

THE CRYSTAL STRUCTURE OF ICE

Why is it that ice plays such a vital role in the energy system of our planet, and indeed of any planet that may contain life? The answer lies in the unique properties of the ice crystal, and these in turn derive from the unique properties of the water molecule which make it the key to life.

An isolated molecule of water, H_2O, has an almost perfect tetrahedral shape, that is, a triangular pyramid (fig. 2.1). In it the electron that normally orbits the proton in the little solar system of a hydrogen atom is instead shared between the proton and the oxygen nucleus, creating what is called a *covalent bond*. There are two such H-O bonds in the molecule and they form a bent geometry in which the angle between the two bonds is 104.5° (a perfect tetrahedron is

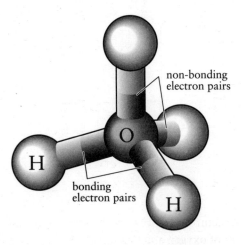

Figure 2.1: Tetrahedral structure of a water molecule.

109.5°). The tetrahedron is completed by two available electron pairs from the oxygen atom, which are not coupled to anything else but which are available for bond formation. What happens when these freely juggling molecules in liquid water freeze into solid ice? We did not know until as recently as 1935, when the three-dimensional structure of solid ice was elucidated by the great chemist Linus Pauling.[1]

The basic building block of ice is the tetrahedron inherited from the free water molecule. Each oxygen atom is at the centre of a tetrahedron bonded to four other oxygen atoms at the vertices separated by 0.276 nanometres (nm, 10^{-9} metres). These oxygen atoms are concentrated close to a series of parallel planes that are known as the basal planes. The principal axis, or c-axis, of the crystal unit cell lies perpendicular to the basal plane and the whole structure looks much like a beehive or honeycomb, composed of layers of slightly puckered hexagons (fig. 2.2).

This structure causes ice to be anisotropic, that is, to have properties which vary in different directions. When an ice crystal grows from the freezing of water molecules onto it, it is energetically easier for new oxygen atoms to be added to an existing sheet of the beehive than to

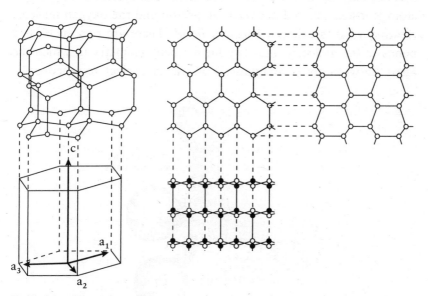

Figure 2.2: The structure of an ice crystal, showing the puckered honeycomb arrangement of oxygen and hydrogen atoms. The c-axis is the axis of symmetry and the other three make up the basal plane of the crystal.

begin a whole new plane, because only two new bonds need to be created instead of four. Ice crystals therefore grow more readily along the axes in the basal plane than along the c-axis – the existing sheets in the beehive get bigger in preference to starting a new sheet. These preferred growth directions turn out to be the directions of the arms of snowflakes growing from vapour in clouds, and of the arms of delicate ice crystals growing on the surface of a newly freezing sea or lake surface. For the purposes of understanding sea ice, the most important thing is that one of them is the preferred growth direction of sea ice crystals in a sheet which is growing thicker by freezing.

We can see these preferred growth directions most easily when a thin layer of water on a windowpane freezes to form a delicate tracery of ice. The first ice crystal to form sends arms shooting out across the glass at angles of 60° to one another, then fills in the gaps with new arms like the branches of a tree. In every case the angle is 60° and the growth of the arms is very fast – this is called *dendritic growth*, from the Greek word for tree.

This is the structure for ice at temperatures and pressures that occur on the Earth's surface; at very high pressures and temperatures near absolute zero (–273.16°C) other, more closely packed forms of ice exist – in fact there are seventeen such so-called polymorphs known to science.[2] The one that is familiar to us, found in normal conditions on Earth, is called *ice 1h*. Some of the high-pressure forms probably exist deep inside planets that are a long way from their suns, and we can re-create them in the laboratory. Other forms exist at temperatures near to absolute zero. They are responsible for some very special processes in outer space. For instance, ice forms the outer part of most comets, and coats the tiny grains of dust in space which cause the stars to twinkle even when seen from above the Earth's atmosphere by an astronaut. It was suggested by the astronomer Fred Hoyle that life may have originated in space on such tiny grains, which would form a substrate holding molecules close enough together for chemical reactions to take place which eventually lead to life. The recent wonderful excursion of the European Space Agency (ESA) spacecraft Philae to the comet 67P/Churyumov-Gerasimenko has shown ice vapourizing and flying out into space in little jets as the comet approaches the sun and its ice cover warms up.

The net of oxygen atoms is held together by *hydrogen bonds*, a type of bond in which an H atom links two Os. Each bond has an H atom lying between the two Os, but the position of each H atom in the bond has to be closer to one O than to the other, with the decision being random as to which to prefer. Each O has two H atoms near it, but there can be only one H along each bond. Subject to these two rules, which derive from quantum mechanics, the hydrogen atoms can be arranged in any way. It is the length of the hydrogen bond that creates the open structure of ice; when ice melts, some of the hydrogen bonds are broken, causing a collapse into a disordered structure of jumbled, random H_2O molecules with a higher density than in the solid state. This makes water very unusual among molecules in that the solid form is less dense than the liquid, unlike, for instance, metals. The density of pure water is 1,000 kilograms per cubic metre (kg m^{-3}) – this was the original definition of the kilogram – and that of pure ice is 917.4 kg m^{-3}. Sea water has a higher density than pure water, typically 1,025 kg m^{-3}, so in the sea the density difference between water and ice amounts to about 10 per cent. This is why 10 per cent of the mass of an ice floe or an iceberg protrudes above the sea surface.

One wonders what would happen if the density difference were in the other direction, as in most substances, and ice sank in water. First, lakes, rivers and even the sea would freeze almost solid. As soon as any ice formed at the surface of a mass of water, for instance at the sea surface due to a low air temperature, the ice would sink right to the bottom and build up a layer there. All bottom life would be wiped out and, in the case of a lake, the ice layer at the bottom would thicken until there was only a small layer of unfrozen water at the surface by the end of winter, or perhaps none at all, in which case all life there would be wiped out. The same would happen to the sea, although it is not clear if winter would be long enough to build up a layer of ice on the ocean bed which would thicken enough to fill the ocean. Certainly growth would be rapid: in our real world the sea ice forms a thin layer at the surface which protects the ocean from further freezing, but in our alternative world the ocean would absorb cold from the atmosphere unrestrainedly through the winter, forming a thickening layer of ice on the seabed. I don't think that anyone has modelled whether such an ocean would freeze up to the surface, but if it did all life in it would cease, except perhaps for tiny organisms. Ocean life would be confined to a band of ocean near the equator where freezing cannot happen; at higher latitudes we would have only solid ice masses extending down to the seabed.

A few other things would be different as well. In the real world, when water freezes it expands, so water in a crack, e.g. in the road or in a rock, expands when it freezes and creates frost damage by cracking the surrounding material. This would stop in our alternative world. Also, skating would be impossible if ice were more dense than water. In the real world the intense pressure of a skate on the ice surface lowers the melting point and the ice just under the skate melts, lubricating it. If water were less dense than ice, pressure on ice would raise the melting point and skating would be impossible.

FREEZING AND MELTING

Let's return to the real world and look at very cold water. We normally think of liquids as having no structure and being composed of random molecules swirling and tumbling around one another. But cold liquid water contains some of the short-range order of ice, with the crystal-like bonded structure remaining within groups of molecules for seconds or minutes at a time until it is destroyed by thermal motion. It is like a group of people at a busy railway station trying to stand together and talk but being split apart by eddying crowds. This accounts for the curious density behaviour of fresh water, which has its maximum density at 4°C. This means that if a river or lake in high latitudes is cooled down in autumn by cold air temperatures, the surface water cools and initially sinks (usually warmer water is less dense than cooler water), to be replaced by warmer water from deeper down, a process called *convective overturning*. This continues until all the water in the lake is cooled to 4°C. Beyond this point, however, the surface water, as it is cooled further, becomes less dense and stays at the surface, so convection ceases. The surface water can then cool down quickly to 0°C and freeze, while the deeper parts of the lake remain near 4°C. So a lake surface freezes over quickly in autumn but it takes much longer for the lake to freeze to the bottom, and in most cases winter ends before this happens.

Sea water does not have this temperature of maximum density; as it cools the water grows denser all the way to the freezing point. The transition from fresh water to sea water behaviour occurs when the salinity exceeds 24.7 parts per thousand (ppt) of dissolved salt in the water; most sea water has a salinity of 32–35 ppt, and it is only a few isolated seas like the Baltic, and regions near the mouths of big Arctic rivers, that have a lower salinity than 24.7 ppt. The loose English term 'brackish' – used for water that is somewhat salty but not as salty as the sea – has a rigorous definition in oceanography, in that it is applied to water that has a salinity of less than 24.7 ppt and so possesses a temperature of maximum density. This means that when proper sea water is cooled in autumn, convective overturning goes on until all the water reaches the freezing point. The freezing point itself is below 0°C since it is depressed by the presence of salt down to –1.8°C for typical sea water (freezing point depression is the main reason why salt is spread on freezing roads). The only thing that prevents the entire depth of the ocean from having to cool before any freezing can occur at the ocean surface is the fact that the ocean is made of layers of different types of water from different sources, all moving in different directions at different speeds. There is a rapid density change (called a *pycnocline*) between each layer, so in practice the convection only has to extend to the bottom of the surface layer – in the Arctic this is called Polar Surface Water, while the layer below it is called Atlantic Water since it reaches the Arctic from the Atlantic Ocean.

The fact that ice floats in water means that sea ice forms a thin floating cover on the sea surface, permitting ocean circulation to go on underneath it and life to dwell both in the deep ocean and, especially, near or even inside the sea ice, where plant plankton (called phytoplankton) have access to the light that they need for photosynthesis. The lower layer of the Antarctic sea ice, for instance, carries tiny liquid brine channels that contain plankton, which are responsible for about 30 per cent of the annual biological production in the entire Antarctic Ocean.

Another key special property of ice is its extraordinarily high latent heat of fusion, 80 kilocalories per kilogram (kcal kg^{-1}). Latent heat is the amount of heat that you have to supply to melt a kilogram of ice when it is already at the melting point, as opposed to specific heat, which is the heat needed to raise the temperature of a kilogram of a substance by 1°C. The specific heat of water is only 1 kcal kg^{-1} – this was the basis for the original definition of the calorie, the standard unit of heat, which is the heat needed to raise the temperature of 1 gram of water by 1°C (so water has been used as the basis for defining two major physical units, the kilogram and the calorie). But if you have to supply only a kilocalorie of heat to warm up a kilogram of water by 1°C, to melt a kilogram of ice (its latent heat) you have to supply 80 kcal, which is the amount that would heat up the same mass of cold water to 80°C. This is a vital contrast. If you put two saucepans together on the stove with equal heat, one containing a kilogram of ice at the melting point and the other containing a kilogram of water at 20°C, room temperature, then the water in the 20°C pan will start to boil at the same time as the last of the ice is melted from the ice pan.

In planetary terms, the latent heat of fusion of water acts as a huge heat reservoir, a kind of buffer to climate change. A prime example is sea ice in summer: it is melting but so long as it doesn't melt away altogether it holds both the air temperature near the surface to about 0°C (since warmer air would melt more of the ice and would be itself cooled in the process), and the water temperature under the ice to about 0°C (warmer water would melt more of the ice and be cooled itself in the process). So long as it continues to exist, the sea ice provides what is effectively an air- and water-conditioning system for the summer ocean.

THE FORMATION OF ICE IN THE SEA

The ice which concerns us most in this book is the ice which forms on the ocean, sea ice. Let us look at how it is created and grows, given what we now know about the special properties of the ice molecule and crystal. We begin by considering calm water freezing in quiet conditions, with no waves present. As the cold atmosphere extracts heat from the water surface, the surface molecules begin to freeze. This produces a skim of separate ice crystals, which initially are in the form of tiny discs or stars, floating flat on the surface and 2–3 mm in diameter. Each disc or star is a crystal with its c-axis vertical, and the disc grows dendritically outwards along the surface (i.e. shooting outwards in six directions at 60° to one another), expanding its beehive sheets into a six-fold snowflake-like shape. The arms of the flat crystals, however, are very fragile, and soon break off, leaving a mixture of discs and arm fragments. These randomly shaped pieces of crystal form a suspension of increasing density in the surface water, like a white slurry, or the medicine known as Milk of Magnesia. This first ice type is called *frazil* or *grease ice*. In quiet conditions the frazil crystals eventually freeze together to form a continuous thin sheet of young ice; in its early stages it is called *nilas*. When only a few centimetres thick this is transparent (dark nilas), but as the ice grows thicker the nilas takes on a grey and finally a white appearance and it is no longer possible to see through it. Once nilas has formed, the sea is separated physically from the atmosphere, and so a quite different growth process begins, in which water molecules freeze on to the bottom of the existing ice sheet, a process called *congelation growth*. This process leads, after further growth, to *first-year ice*, which in a single season in the Arctic reaches a thickness of about 1.5 metres and in the Antarctic 0.5–1 metre. In the Antarctic, and in areas with plenty of waves and turbulence, the frazil stage can go on much longer and plays an important climatic role (see chapters 11 and 12).

Once a continuous sheet of nilas has formed, the individual crystals which are in contact with the ice–water interface grow downwards by the freezing of water molecules on to the crystal face. This freezing process is easier for crystals with horizontal c-axes than for those with vertical c-axes, because the downward growth can be achieved by extending existing beehive sheets. So crystals with horizontal c-axes grow at the expense of the others, and, as the ice sheet grows thicker, crowd them out in a form of crystalline Darwinism. After about 20 cm of growth the selection process is complete and the favoured crystals continue to grow downwards, creating a fabric composed of long, vertical, columnar crystals with horizontal c-axes. This columnar structure is a striking feature of first-year ice even when viewed with the naked eye. You can also see that such an ice sheet is likely to be mechanically weak, because it is essentially a bundle of crystals all oriented in the same direction.

What happens to the salt that was dissolved in the sea water? The crystal structure of ice is a very open one, but it is not so open that other molecules or atoms can easily incorporate themselves in the spaces within it. When ice grows from salt water, therefore, the salt molecules cannot enter the crystal structure. However, the salt does get into the ice in a different way. The advancing ice–water interface is not flat but composed of parallel rows of projections called *dendrites*, each of these representing a few sheets of rapidly advancing (i.e. dendritically growing) beehive, with narrow, water-filled grooves between them. From time to time bridges of ice grow between successive projections, trapping water in the intervening groove in the form of an isolated cell, called a *brine cell* (fig. 2.3).[3] Quickly the walls of the cell freeze, eating into the cell volume until all that is left is a tiny blob, about half a millimetre across, of very concentrated brine solution which does not freeze. These brine cells contain the salt which makes first-year ice still taste salty (young ice contains about 10 ppt of salt, compared to about 32 ppt for the parent sea water). This brine slowly drains out of the ice through the winter, by various mechanisms such as brine cell migration, brine expulsion and simple gravity drainage. Brine cell migration occurs because the top of each cell is at a slightly lower temperature than the bottom, since in winter there is a steep temperature gradient between the ice–water boundary at $-1.8°C$ and the ice–air boundary at more like $-30°C$. The top of the cell freezes, the remaining water in the cell becomes more saline, and the bottom melts; the whole cell moves downwards in the ice sheet, carrying its brine with it. Brine expulsion occurs when the whole cell is trying to freeze as the temperature drops; pressure builds up in the surviving tiny brine-rich droplet and it explodes, forcing brine out and downwards. Gravity drainage, the most efficient process, works because, as the ice grows thicker from freezing below, existing brine cells are lifted above the waterline so that gravity encourages the brine to find paths through interconnecting pores so as to drain out of the bottom of the ice. These paths tend to coalesce, like river tributaries, into channels called *brine drainage channels*. When summer comes, all the surface snow on top of the ice melts and so does some of the ice itself. Fresh water gathers in melt water pools on the surface, and the melt water works its way through the ice, flushing out most of the remaining brine, a process called, obviously, *flushing*. If the ice then survives the summer into a second year of growth it is now almost fresh, does not taste of salt, and is much stronger – this type of ice, called multi-year ice, has always been a more formidable obstacle for icebreakers than first-year ice.

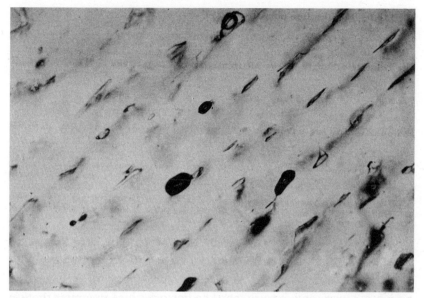

Figure 2.3: Tiny brine cells in sea ice. Spacing between brine layers is 0.6 mm.

THE IMPORTANCE OF SUMMER MELT

We will see that the process of forming melt pools is very important from a climate change point of view. In winter, when a fresh snow layer covers the sea ice, the surface reflects 80–90 per cent of the solar radiation falling on it, so we say its albedo (its reflectivity) is 0.8 to 0.9. When the snow melts and one is left with bare ice, which may have some coverage of dirt from the black carbon (soot from the atmosphere) which has accumulated on the snow during the winter, the albedo drops to 0.4–0.7; this occurs in June–July, just when solar radiation is at its peak with 24 hours of daylight and the sun high in the sky. If this surface of bare ice and melt pools were to be created even slightly earlier in the summer, the additional radiation absorbed would play a major role in thinning the ice and would perhaps allow it to melt altogether. This, many Arctic scientists think, is what is now happening, helping to cause an irreversible loss of summer ice.

As the melt pools grow deeper and wider they may eventually drain off into the sea, over the side of floes, through existing cracks, or by melting a *thaw hole* right through the ice at its thinnest point or at the melt pool's deepest point. The drained water forms a low salinity layer a few metres deep which bathes the underside of the sea ice and enhances the rate of bottom melt.

HOW ICE FORMS LEADS AND PRESSURE RIDGES

So far we have considered how sea ice forms and changes under thermal processes alone, growing and melting on the sea surface. Yet in the Arctic only about half of the ice volume has formed in this way, the rest having formed by the deformation of existing ice, with ice being piled up into linear *pressure ridges*, and openings being created by this process which are called *leads*. Here is how it happens. The pack ice sheets that have formed by freezing and growth are constantly in motion, driven by the frictional stress of the wind on their upper surface and the water currents on their lower. This creates overall patterns of surface drift based on the prevailing winds. In the Arctic, for instance, there is a wheeling current system which rotates clockwise in the North American side of the Arctic Basin, known as the Beaufort Gyre, while north of Europe ice is gathered from Siberian waters and moved by the wind across the Pole and down to Greenland, a current known as the Trans Polar Drift Stream.

The wind stress which drives this sea ice is integrated over a large area; it has been estimated that in concentrated pack ice a piece of sea ice responds to winds integrated over a distance of 400 km upwind. Therefore if the wind varies over a large area, it may create what is known as a divergent wind field and thus a *divergent stress* – that is, the pattern of wind is acting to tear the ice cover apart. Since ice has little strength under tension, this divergence can open up cracks, which widen to form *leads* (Plate 6). In winter any leads formed in this way rapidly refreeze because of the enormous temperature difference between the atmosphere (−30°C typically) and the ocean (−1.8°C). The heat loss from a newly opened lead can be so violent (more than 1,000 watts per square metre, $W\ m^{-2}$) that the lead steams with *frost smoke* (Plate 5) from the evaporation of the surface water exposed by the lead opening. Naturally a young ice cover rapidly forms to heal this opening within hours, by creating nilas, and this cuts out the evaporation. When a subsequent wind stress field becomes *convergent* – that is, it pushes the edges of the floe together – the young ice in the refrozen leads forms the weakest part of the ice cover and is the first part to be crushed, building up heaps of broken ice blocks above and below the water line. Such a linear deformation feature (it is like a long slag heap) is called a *pressure ridge* (Plate 7), the above-water part being called the *sail* and the (more extensive) below-water part being called the *keel*. Keels in the Arctic can be as deep as 50 metres, although most are about 10–25 metres deep, with depths of 30 metres seen only every 100 km or so of track. The keel is typically about four times deeper than the height of the sail, and is also two or three times wider, so that apparently undeformed ice near a pressure ridge may have part of the keel underneath it; this occurs because it is easier to push ice blocks downwards against buoyancy than upwards against gravity.

Ridges in the Arctic make a major contribution to the overall mass of sea ice; probably about 40 per cent on average and more than 60 per cent in coastal regions. Ridges start off as simple linear piles of ice blocks, but gradually grow stronger as the ice blocks freeze together, so that after a few years a ridge, like a healed scar, can have a strength equal to or greater than the surrounding undeformed ice. It is the heavy consolidated ridges in multi-year ice which make that ice so impassable to any except the heaviest icebreakers. First-year ice, however, is not only thinner in the first place but also includes ridges which have not had time to bond together strongly in this way, so the ice is much weaker and offers less of a barrier to an ice-strengthened ship.

In the Antarctic, ridges are much shallower than in the Arctic, usually less than 6 metres in draft. The reason is that the ice itself reaches a lesser thickness after a year's growth than in the Arctic, only 0.5–1 metre compared to 1.5 metres. These thin sheets can easily be buckled directly by the stress of the wind, without the need to form and crush a lead first. So the block thickness in the ridge is often similar to the thickness of the floes on either side, and there is no opportunity to build up a high ridge by the progressive crushing of a refrozen lead. The ridged ice also seems to contribute less to the overall volume of ice, maybe only 30–40 per cent. For a further discussion of Antarctic sea ice, see Chapter 12.

ICE IN SHALLOW WATER

When ice first forms on the sea, very often it is on the shallowest water near the beach, because the atmosphere only has to cool a thin layer of water to allow the top surface to freeze. This is called *landfast ice*, or *fast ice*, because it is frozen to the seabed. A little further out, beyond one or more tidal cracks, the ice is afloat but remains stationary because it is pinned to features which are themselves aground. These are usually pressure ridges which have been driven inshore by the wind acting on the freely floating pack ice, and which run aground in the shallower water. Young ice grows around these grounded ridges and the entire region is called the *fast ice zone*, which extends out to a water depth equal to that of the deepest grounded ridges, usually 25–30 metres.

While the offshore ice is still in motion, before it is brought to a complete halt by running aground, the crests of the grounded ridges excavate long narrow troughs in the seabed sediments, a process called *ice scour*. Ice scouring was discovered during my first voyage to the Arctic aboard the *Hudson* in summer 1970. A team from the Geological Survey of Canada was towing a sidescan sonar behind the ship, which sends out a fan-shaped acoustic beam to map the seabed and give reflections from any obstacle. We all expected the map of the muddy nearshore zone to be a blank – a great expanse of featureless silt. Instead we saw a complex array of long narrow troughs in the seabed, as if a drunken ploughman had been at work. It was a fascinating pattern of intersecting lines, some straight as a die and others curved round into circles and spirals, like a raked Zen garden in Japan. Old lines were over-written by new lines. I remember rushing up to the main echo sounder and finding that each scour mark that crossed the ship's track showed up as a small indentation in the seabed, 2–4 metres deep. We realized straight away that these scour marks must have been made by pressure ridges embedded in the winter ice cover and dragged over the bottom by the force of wind and current on the pack before being brought to a halt. The pressure ridge keel with its mountainous shape acts as a multiple plough. The scour marks pose a previously unsuspected hazard to any plans for offshore pipelines or wellheads in shallow Arctic waters.

More complete surveys of ice scouring have shown that in places it extends out to water depths beyond those where ridges can form, sometimes to 65 metres of water (when, as I have said, pressure ridges seldom exceed 30 metres). The explanation is thought to be that these are ancient scours dating from during or just after the last Ice Age, when sea level was lower because of the water locked up in ice sheets. The very slow rate of sediment deposition in Arctic waters, because of the dearth of plankton (whose tiny shells rain down on the sea-bed), means that these ancient scours have survived to the present day without being filled in.

During the 1970s, when scientists extended sidescan sonar surveys into deeper water, they discovered *iceberg scours* in water depths of 150–300 metres in the Labrador Sea, Baffin Bay, off Greenland and in the Antarctic, all places where the deepest underwater peaks of a drifting iceberg had ploughed through the seabed. Amazingly, this also provided the very first evidence that Mars once possessed flowing water. My friend and colleague Chris Woodworth-Lynas, who lives and works in Newfoundland, is an expert on iceberg scours and has also found them on land, in places like King William Island in the Canadian Arctic (where Franklin's men perished). This island was sea-bed during the last Ice Age and icebergs from nearby glaciers traced out curving patterns of scour into the boulder-strewn sediments which were later raised up to form the present visible surface of the island. In 2003 Chris was browsing through photographs of the surface of Mars, obtained using the Mars Orbiter Camera on the Voyager space-craft, and saw an exactly similar pattern.[4] His paper with his colleague Jacques Guigné was a breakthrough in Mars studies. Today we read-ily accept that Mars once had water and therefore, perhaps, life, but in 2003 this was a heretical view. The scour marks showed that not only did Mars once have flowing water, but that this water was peri-odically frozen (perhaps during the winter season only) to form icebergs or ridges which scoured the ancient Martian seabed.

The processes that go on in shallow water are very complex. Beyond the fast ice the frictional drag of the stationary ice on the rapidly moving offshore pack slows it down and produces what is known as the *shear zone*, a region where the friction and pressure can create deep ridges, sometimes in a huge region of confused broken blocks called a *rubble field*. One fascinating product of this process is a form of enormous isolated ridge, known by its Russian name of *stamukha* (plural stamukhi). A stamukha, found typically in the shallow waters north of Siberia, is a deep ridge which grounds during the winter and becomes part of the fast ice zone, but which does not lift off and drift away in spring or summer because it is too firmly aground. The ice all around it breaks up, leaving it as an isolated domed island of ice in a vast expanse of open water. It may be very dirty and so look like a real island, because in early spring the melt water from Siberian rivers will have discharged mud all over it. Eventually it breaks away from the seabed and drifts off into the Arctic Ocean, where it forms one of the most formidable obstacles to ships or drilling rigs. Stamukhi are very rare in the pack ice, but I was fortunate in finding and examining one in the summer of 2012 in Fram Strait, between Spitsbergen and Greenland. Plate 9 shows the great humped surface of the stamukha, with a russet surface due to a mixture of many years of dirt and algae. I sent an AUV (autonomous underwater vehicle) under the ice to map its draft using multibeam sonar. It had a draft of 28 metres, which is more than enough for it to have run aground in a typical shear zone.

POLYNYAS

Finally, there are places along polar coasts where instead of fast ice or piled-up ridges there is open water, even in winter. These areas are called by the English plural of a Russian name, *polynyas*, meaning pools. They can be formed in many different ways, but a typical one is the presence of a prevailing offshore wind. The wind blows new ice out to sea as fast as it forms, leaving a stretch of open water against the coast which may extend tens of kilometres out to sea. In winter the open water near the shore steams with frost smoke from evaporation, while further out ice is forming as frazil ice and is being driven further out by the wind until it meets the heavier offshore pack. The coastline of Antarctica is ringed by a succession of polynyas because of katabatic winds: these are winds which accelerate as they blow down the slopes of the domed Antarctic ice sheet and out to sea, in concentrated form, through the gaps between coastal mountains. Each gap, which usually corresponds to a glacier, gives rise to its own polynya. The polynyas are recurrent and usually have names. Plate 11 shows the location of the Terra Nova Bay polynya in the Ross Sea, where the Italians and now the South Koreans have bases, and where Captain Scott's northern party was forced to spend a winter living in an ice hole.

In the Arctic polynyas are less common, but can be very important. St Lawrence Island in the Bering Sea has a polynya on its south side because of prevalent northerly winds in winter, and this allows the local Inuit to hunt and fish through the winter. There is a famous polynya between northwest Greenland and Ellesmere Island called the North Water (Plate 12); this forms in a different way because the wind and current drive ice southward down through a narrowing opening between the two great islands, and it can get stuck, forming an arching barrier like wet sand trapped in a hopper. The water carries on southward while the ice stays behind, creating a winter polynya. Another recurring polynya is called the Northeast Water and is found on the northeast coast of Greenland where the southward flowing ice in the Arctic pack cannot 'turn the corner' fast enough at the southern end of the protruding Nordostrundingen promontory, leaving an area of open water in its lee. It is here that Danish archaeologists have discovered an ancient umiak (an open boat made from skins) and stone tools, showing that perhaps a thousand years ago the Inuit had a hunting settlement in this utterly remote far-northern outpost, at 81°26'N, probably because of the large populations of polar bears and seals to be found here.

This chapter has briefly summarized the properties of sea ice and how it forms and grows on the sea surface. We will see that this fascinating material is one of the most important substances on the planet, because of the serious climate impacts associated with its retreat. But before dealing with these we shall take a look at the other kind of ice encountered on the Earth's surface, the solid pure ice of glaciers and ice sheets. Though more slowly than sea ice, this too is disappearing.

3

A Brief History of Ice on
Planet Earth

THE FIRST APPEARANCE OF ICE

We do not know when or how water in its frozen form first appeared on the surface of planet Earth. When, 4.54 billion years ago, the Earth first condensed out of the solar nebula, the rotating disc of gas and dust that had accumulated around our Sun, it was an extremely hot young planet. In fact the surface of the Earth was molten, partly from volcanic activity and partly from frequent collisions with the mass of dust and rocks that still remained in the solar nebula. The atmosphere would have been made up of toxic gases with almost no oxygen. No form of life that we know about could have lived in such inhospitable surroundings. Nevertheless, life of some kind did begin, as early as 3.8 billion years ago (some scientists say 4.1 billion), and although it does seem that the Earth's surface had solidified by then, and may even have carried some liquid water, there was certainly no ice. Interestingly, the oldest known fossil, which consists of some graphite that was determined to have had a living origin, was found in some 3.76-billion-year-old rocks in West Greenland. Of course, it wasn't Greenland then, but a liquid ocean under which these primitive organisms lived in mud layers.

The fascinating story about life, right from the very first, is that everything we know tells us that liquid water is essential to its generation and continuation. Water is the main component of the living cell. But where did the cooling Earth's first water come from? It is assumed to be a combination of out-gassing from the Earth's interior and impacts from comets and asteroids that were made largely of ice.

These early forms of life were tiny, microscopic single-cell organisms, and remained that way until about 580 million years ago. So about 80 per cent of the history of life on Earth consisted of very slow changes to single-celled organisms. Then, suddenly, multi-cellular life appeared, and evolution took off because there were now infinite possibilities for combinations of cells, specialization into different roles, and then evolution of organs and limbs. We haven't looked back since. It is extraordinary that this critical step from one cell to multiple cells took so long, because as early as 2 billion years ago the first organisms had appeared that were capable of photosynthesis – absorbing solar radiation and, with the aid of carbon dioxide, creating new body mass, accompanied by the release of oxygen. From that date onwards the Earth's atmosphere started to become more oxygenated and so more hospitable to forms of life that we know, like single-celled plants, in the ocean and on land.

This whole vast period, from the formation of the Earth to the formation of the first multi-celled organisms, is called by palaeontologists the Pre-Cambrian, covering nearly 4 billion years of Earth history. During all that time, where was the ice?

THE PARADOX OF SNOWBALL EARTH

Palaeoclimate researchers have divided Earth's climatic history crudely into periods of a 'hothouse Earth', when the planet was distinctly warmer than today, and 'icehouse Earth' when it was distinctly cooler. About 75 per cent of the time the planet was a hothouse. Yet, curiously, it seems that there were ice ages during the Pre-Cambrian that were far more severe than those of recent time. The first that we know about is called the Huronian glaciation by geologists or, less pronounceably, the Makganyene glaciation, after a location of glacial deposits in South Africa. It occurred from 2.4 to 2.3 billion years ago, before photosynthesis and before there was much oxygen in the atmosphere, when the Earth was a very hostile environment to slowly developing single-celled life. This glaciation was very severe and long-lasting, much more so than recent glaciations, and must have had a traumatic effect on life on Earth.

It was this glaciation which led some scientists to develop the idea that the entire planet became frozen over, both its ocean and land surfaces, so that the Earth had a very high albedo and looked white from space. This was termed 'Snowball Earth' and has been a controversial concept since it was first introduced by Joseph Kirschvink of the California Institute of Technology in 1992.[1] The concept is now quite solidly established, though not universally accepted. The questions that we have to answer are: how did it start? How long did it last? How did the Earth emerge from it? And what did the planet look like while it was a snowball? It is hard to be sure of exact answers to these questions, since evidence of such ancient processes is hard to find, but the story probably goes something like this.

At the time of the Makganyene glaciation the Sun was less luminous than it is now. Today we are used to thinking of the 'solar constant', the amount of radiation reaching the Earth from the sun, averaged over the year, and to consider this a genuine constant with only tiny variations (though, according to a few people, these variations are enough to cause climate change). But today's solar constant is the result of a very slow but steady rise in the sun's luminosity, which has increased by about 6 per cent every billion years. With the sun about 15 per cent dimmer than today, the Earth of 2.3 billion years ago was kept warm – in fact warmer than today – by the huge amounts of carbon dioxide and other greenhouse gases, such as methane, emitted by vigorous volcanic activity. If such activity had suddenly slowed down, the whole Earth would easily have sunk into a much colder state than today. The glacial deposits found in South Africa, which give their name to this glaciation, were laid down when South Africa was near the Equator, implying that the glaciation was worldwide, so the concept of a slowing of volcanic activity could well be the mechanism. Of course, today we can have glaciers at the Equator – on the top of Mount Kilimanjaro, for instance – but the Makganyene were at low altitude, suggesting a totally glaciated world.

Another explanation for this ice age was that in fact the first photosynthetic organisms were already evolving and beginning to change the atmosphere, producing more oxygen. Oxygen reacts with methane to form carbon dioxide, itself a greenhouse gas but one much less effective (by a factor of 23 per molecule) than methane. So the copious amounts of methane in the atmosphere would have been diminished by oxidation as soon as this free oxygen started to appear. We have a Gaia-like case of life modifying the environment of the planet by cleansing the atmosphere of methane and thus cooling it down, though the resulting ice age can scarcely be described as benign towards life.

What would life be like on Snowball Earth and how long did the snowball last? To some extent it would be self-sustaining. The snow and ice covering the surface would have raised the average planetary albedo to something like 0.8 (its present value is 0.3), so most incoming solar radiation would have been reflected back into space. As a result, it has been calculated that the average temperature of Snowball Earth might have been about −50°C, with the warmest point being the Equator at −20°C, inhospitable indeed. One of the unknowns is how thick the ice became over the oceans. It is likely that the thickness was very great, perhaps 1 km, resembling the floating ice shelves of Antarctica today although formed at sea instead of on land. The thickness would also have been less at the Equator because of the warmer temperature, so the thicker ice at higher latitudes would have flowed, like a present-day glacier, towards the Equator. Yet this flow was in an ice sheet floating in the ocean, not an ice sheet on land. And at the centre, at the Equator itself, estimates of thickness range from hundreds of metres down to just 1 m. There is a big difference. If it were 1 m there would be plenty of cracks and open leads where gas and heat could be exchanged between ocean and atmosphere and where, crucially, photosynthesis by ocean organisms could continue, building up the planet's oxygen levels. Yet volcanism would also have continued, through underwater eruptions along mid-ocean ridges and emissions of gases through thermal vents, so putting carbon dioxide and methane back into the atmosphere. By one or other of these methods the gases would have built up to the point where the greenhouse effect was sufficient to melt the ice sheets and take the Earth back into a warm state. This may have happened very quickly, in perhaps only 2,000 years after millions of years as a snowball. Single-celled life in its myriad forms is extremely resilient, and many species would have survived the lengthy period of cold.

TWO MORE SNOWBALLS

The story of the first snowball is not complete and is the subject of disagreement among geologists and climate modellers. But it seems that this state of the Earth reappeared, after a huge gap of 1.5 billion years, in the form of the Sturtian glaciation, a mere 710 million years ago. This time the mechanism is hypothesized to be carbon dioxide. The Earth has always been subject to the process of plate tectonics, whereby both continental and oceanic crustal rocks are moved around in the form of plates which interact along their edges by over-riding one another, while fresh crust is created elsewhere from rising fluid material originating from the Earth's mantle. It so happened that 710 million years ago the plates carrying the land masses had moved together to form a single giant land mass, Pangea, which was concentrated at and near the Equator. This accelerated a process known as silicate weathering, whereby magnesium silicate in rocks reacts with carbon dioxide to form bicarbonate and silicic acid in solution (this has recently been proposed as a way of reducing CO_2 levels in the atmosphere, by breaking up silicate rocks into small fragments and spreading them on beaches – see Chapter 13). The exposed silicate rocks were warm, speeding up the process. They were about to become wet, too, because Pangea began breaking up into smaller fragments, which moved apart to become new continents, producing more coastal terrain, with rocks weathered by rain, and less interior terrain, which was often desert. Between the two effects it seems that carbon dioxide was consumed by the weathering process, allowing the Earth to cool and a new snowball to set in, which lasted about 60 million years.

Finally, 635 million years ago, the third and final snowball seems to have occurred, called the Marinoan glaciation and following soon after the end of the Sturtian (both of these glaciations are named after locations in South Australia). This lasted for between 6 million and 12 million years. Again carbon dioxide and weathering are believed to be involved, but many other mechanisms have been invoked, including astronomical ones such as some massive cloud of space debris blocking out solar radiation.

The concept of a Snowball Earth is such a new one, and the evidence so difficult to find after such vast lapses of time, that it is possible that the whole idea will in the end prove to be invalid. This doesn't do away, however, with the fact that during the Pre-Cambrian era there were three, and as far as we know only three, huge long-lived glaciations, even if they did not cause a snowball. This means that the natural state of the Earth during that immense period of time was to be unglaciated, and in fact often warmer than today, with the very occasional glaciations representing sudden breakdowns of the global thermostat. What happened since then to bring us into the world of endless repetitive ice ages that we have had for the past 6 million years?

THE TRANSITION TO ASTRONOMICALLY DRIVEN ICE AGES

Through most of the past 600 million years of planetary evolution we have had a 'hothouse Earth', as opposed to an 'icehouse Earth'. Palaeoclimate research is such a young science, and so little of the Earth has been properly examined for evidence of past dramatic changes, that we cannot be sure that there have not been many ice ages, maybe short-lived ones, over that period. The picture of Earth history is still very crude, but it does look on the whole as if the Earth has been warm during most of this time.

This is not a book about geology, nor does it try to look at the whole history of climate change on our planet. My primary concern is to look at ice and its role, and to see what the implications are of that role coming to an end. But to do that we do need to look at other periods in Earth's history when carbon dioxide levels were rising very fast, and draw some lessons about what that means for the planet. One of the lessons from climatic history is that there is no period in Earth history that we know about where the rate of rise of atmospheric CO_2 is as great as it is today. Human beings are truly carrying out a global experiment involving an unprecedented level of interference with the natural system.

A major natural event occurred 65 million years ago, when the famous K-T asteroid impact on the Yucatan Peninsula in Mexico

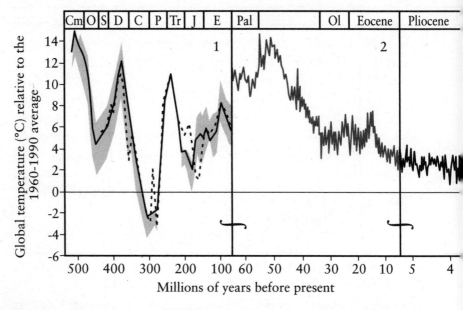

Figure 3.1: Temperature record of Planet Earth. Curve 1 is warm temperature calculated from isotope ratios in deep sea sediments, shows stepwise tempera-present-day temperatures, reached about 3 million years ago. Curve 4 shows

produced a global disaster. The shock wave and tidal waves travelled round the Earth, and the massive quantity of soil, rock and dust blasted into the atmosphere brought darkness and death. There must have been a series of very cold winters, like the 'nuclear winter' predicted as the outcome of a nuclear war.[2] The dinosaurs became extinct, because their lack of temperature control made them unable to cope with rapid change. But over a few thousand years, when immediate impacts had subsided, the result was actually a warming. There was a rise of more than 2,000 parts per million (ppm) in CO_2 levels and a temperature rise of 7.5°C over a period of about 10,000 years (a rate of 0.2 ppm per year and 0.00075°C per year). During the catastrophe about 4,500 billion tons (or 4.5 gigatons, Gt, equal to 10^9 tons) of carbon had been released from impacted carbonates and shale, from ignited bushfires and from ocean warming, and this caused the rise. Yet the CO_2 rise rate was still an order of magnitude lower than the current rate of 3 ppm/year. We are injecting greenhouse gases into the atmosphere far faster than any known natural event, even an extreme one like an asteroid impact.

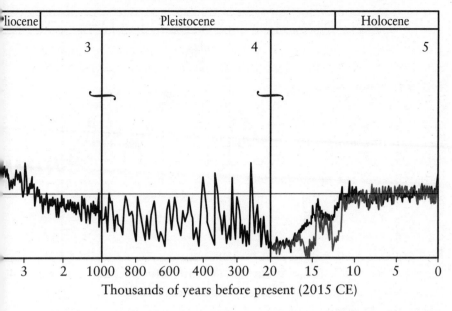

from 500 to 80 million years ago. Curve 2, from 60 to 6 million years and ture drops through Eocene to Pliocene. Curve 3 shows a further drop towards alternating ice ages and curve 5 our recent recovery from the last Ice Age.

Ten million years later, a large-scale release of methane from a source in Siberia (of whose cause we are uncertain) drove a climatic warming in which atmospheric CO_2 also rose to nearly 1,800 ppm with a temperature rise of about 5°C. But this all took 10,000 years so that the growth rates were only 0.18 ppm of CO_2 per year and 0.0005°C/year. This too was an event in Earth history which involved extinctions of species and changes in environment, yet which still involved rates of CO_2 rise that were less than we are imposing on our planet today.

After these very warm episodes the temperature of the Earth gradually declined in a series of steps over 50 million years. This can be seen, for instance, in the temperature record of deep ocean waters (fig. 3.1, curve 2), which can be deduced from the ratio of oxygen-18 to oxygen-16 isotopes in the shells of benthic foraminifera (tiny marine creatures which secrete shells) deposited in sediments. This ratio between normal oxygen-16 and the isotope oxygen-18, which has two extra neutrons, changes as temperature changes in the water in which the foraminifera grow. Scientists are unsure of the reasons for this long-term temperature decline, except that greenhouse gases are probably involved, as is a change in the distribution of the continents over 50 million years, with Antarctica moving to a high latitude and developing an ice sheet.

This finally brings us to the 'modern' era of repetitive ice ages. For about the last 6 million years the average temperature of the Earth has been low enough for small changes in the shape of the Earth's orbit about the Sun to produce small changes in the distribution of radiation over the Earth's surface, so as to change average temperature and bring about the periodic advance and retreat of glaciers – the ice ages. It is extraordinary that it was not until the latter half of the nineteenth century that geologists even acknowledged that *one* Ice Age had occurred in Earth history, known initially as the Great Ice Age. This was an almost embarrassingly recent event, occupying only a few tens of thousands of years and ending about 12,000 years ago. Now we know that this was simply the latest in a periodic sequence of ice ages and warm interglacial periods going back 6 million years. Before that, as described above, the Earth was too warm for ice ages except for a small number of exceptional events (Snowball Earths) occurring in the earliest part of our planetary history and lasting for possibly millions of years at a time. To achieve periodic ice ages the climate has to be cold enough for astronomical fluctuations to cause the advance and retreat of glaciers, but not so cold as to give us a permanent ice age. The sawtooth fluctuation that is the Ice Age temperature history of the Earth is thus probably a modern phenomenon of the last few million years; the alternation of cold and warm periods is not at all a characteristic of most of Earth history. Some scientists believe that the CO_2 and other gases that we are releasing into the atmosphere by fossil fuel burning will be enough to suppress the descent towards the next ice age, in fact possibly enough to suppress the ice age cycle completely so that we will go back to the permanently warm Earth that we had tens of millions of years ago.

Let us now look at this remarkable cycle of ice ages which has governed the climate of our planet in recent time.

4

The Modern Cycle of Ice Ages

THE PLIOCENE AND THE ICE AGES

The entry period to our modern climatic world was the Pliocene (5.3–2.6 million years ago). Global Pliocene temperatures were on average 2–4°C warmer than pre-industrial modern temperatures, and sea level was 25 metres higher. This implies that there was less water tied up in ice sheets, and indeed it seems that Greenland was ice-free although Antarctica had an ice sheet. There was no Arctic sea ice. In fact conditions were rather like those that we are heading into as we modify our own climate, although we do not expect such a huge sea level rise in the short term. The warm temperatures drove an intense hydrological cycle with extreme evaporation and precipitation, which brought extensive rain forests, lush savannahs (where there are now deserts) and small ice caps (about two-thirds of the present area). Early ancestors to Man existed on the Earth but were too few to have an impact on its development. Agriculture would not have been possible because of extreme downpours and heat waves; the idea of planting seeds and expecting them to grow would not have occurred to our ancestors in such an environment. The main thing about this period is that it was cooler than what had come before it, but still too warm for a glacial cycle.

But cooling then continued during the Pliocene. One contributory factor was, until recently, believed to have arisen from the ceaseless drifting of the continental plates, which caused the isthmus of Panama to be created so that North and South America were no longer separated. This destroyed a huge equatorial ocean circulation and prevented warm Pacific water from reaching the Atlantic, which therefore grew much colder. However, this idea has recently been demolished by Peter Molnar of the University of Colorado, who has shown that the isthmus was in place 20 million years ago rather than 3 million, so that it cannot in itself have been a major cause of the cooling. We must seek new causes, but the climate record certainly shows that about 3 million years ago, towards the end of the Pliocene, the global climate became cool enough for an ice sheet to form on Greenland, and the stage was set for our modern oscillating ice ages.

THE RECORD OF RECENT GLACIATIONS

At this stage something new happened in the world climate. As we saw in the previous chapter, the Earth had had 2 billion years in which the climate changed slowly, and was generally warmer than it is today, but where on rare occasions a descent occurred into a harsh and long-lived ice age, which may have frozen the entire planet into a 'Snowball Earth'. What we didn't have was an alternation of warm and cold climatic phases over a period of a few tens of thousands of years, but this is just what now started up. We don't know how long it will last – and we may have already destroyed the cycle by our present actions.

We have an excellent record of these climatic changes in the form of *ice cores* from Greenland and the Antarctic ice sheet. As snow falls on an existing ice sheet, it forms a layer which is then covered over by new snow during subsequent years, and so becomes compressed. The thickness of such an annual layer decreases and its density goes up as it is squashed by the new layers on top of it. Freshly fallen snow might have a density of only 0.3 (300 kg m^{-3}), but when it has been pressed down so that it is 50 m below the new ice sheet surface its density will have gone up to 800 kg m^{-3}. As it is compressed its nature changes, from the original light flaky structure of snow to a more granular structure, called *firn*, then finally into ice itself. The official density at which compressed snow turns into ice is 800 kg m^{-3}. Below this density the crystals of firn are separate enough for air, or melt water, to percolate freely between them. At 800 kg m^{-3} the crystals become welded together by pressure and the material becomes a continuum, although the air channels remain to some extent as closed-off air pockets, which gradually shrink as further compression takes place. Deep down inside the ice sheet there are only very tiny air bubbles which are under high pressure. Right at the top of the ice sheet each annual layer is of measurable thickness, and can be clearly seen if a vertical face is revealed, such as the side of a tabular iceberg. Deeper down, the compressed layers become thinner and thinner until individual years cannot be seen any longer and we must calculate age from our knowledge of the compressibility of ice.

So nature creates for us a layered record of ice going back about a million years, which is as far as we can get if we drill down through the ice sheet right to the bedrock. This is why we are not quite sure when the cycle of ice ages started. How can we read this record? Fortunately, there is an excellent way of calculating the temperature at which the snow fell that created the layers. As we saw in the previous chapter, oxygen comes in two types of atom, called isotopes, the 'normal' oxygen-16 (O^{16}), with eight protons and eight neutrons, and the rarer 'heavy' O^{18}, with two extra neutrons. Usually O^{18} is present at about 1 part in 500. When water evaporates from the sea surface the lighter water molecules (H_2O^{16}) evaporate more easily than the heavier molecules (H_2O^{18}). This water vapour with enhanced oxygen-16 levels then condenses into ice crystals in clouds, in which a further fractionation takes place. Experiments have been done to relate the $O^{18} : O^{16}$ ratio in the fallen snow to the air temperature at the time that the snow formed. This is a perfect thermometer for past climates. Scientists are also becoming much more clever at extracting the tiny amounts of air from the highly compressed bubbles in the ice and analysing them for carbon dioxide and methane content. So not only do we know the temperatures through the last million years but also the atmospheric greenhouse gas concentrations. In addition, dust concentrations can tell us something about how dry the climate was, and thus how many deserts the Earth possessed. And, finally, big volcanic eruptions leave layers of ash, which indicate when possibly climate-changing eruptions took place. In a recent study 116 volcanic eruptions have been identified from the last 2,000 years, the largest being a mysterious eruption in 1257 which must have affected climate for two or three years and which was finally traced to the Samalas volcano in Indonesia. The second biggest was in 1458 near Vanuatu, and the third was the explosion of Mount Tambora in 1815, which killed 71,000 people and which gave Europe a 'year without a summer' in 1816 which caused poor harvests and thus huge social unrest immediately after the defeat of Napoleon.[1]

Reading this record began when the first ice cores were drilled through the Greenland and Antarctic ice sheets in the 1950s and 1960s. As time has gone on, techniques of analysis have improved enormously, so whereas at first there was just a crude record of glacial and interglacial periods, there is now a highly detailed temperature and gas record showing a vast number of unexplained temperature excursions together with a marvellously detailed record of how we enter an ice age, how we leave it, and, most amazingly, how similar the last four ice ages have been to one another, as if the Earth has been subjected to a regular oscillation. What could have caused this?

THE ASTRONOMICAL THEORY
OF ICE AGES

The key to the cyclic pattern of the ice ages is ascribed to the Yugoslav scientist Milutin Milankovitch in 1920, though a strong case can also be made for James Croll, a remarkable Scottish scientist who was completely untutored, worked as a caretaker in a scientific library in Glasgow (the Andersonian College and Museum), and acquired a knowledge of science by reading the books there. He suggested this mechanism in 1867 and his work in this and other fields is only just beginning to be recognised.[2]

The idea is that although the total amount of radiation reaching the Earth over a year remains essentially the same (the *solar constant*), its distribution with season and with latitude varies because of three types of oscillation in the Earth's orbit (fig. 4.1).

The first is the variation in the eccentricity of the ellipse which constitutes the Earth's orbit round the Sun. This orbit is nearly, but

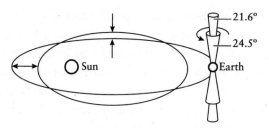

Figure 4.1: The three types of oscillation of the Earth's orbit.

not quite, a circle. When it is closest to a circle, the same amount of solar radiation falls on the planet all year round, but when the eccentricity is greatest the radiation goes through a distinct maximum and minimum each year. At the moment the eccentricity is 0.0167, which means that the semi-major axis of the orbit has a length of 1.0167 times the average distance of the Earth from the Sun, while the minimum distance is 0.9833 times the average. The period needed for the Earth's orbit to accomplish one cycle of shape-changing, from maximum eccentricity to minimum and back again, is 100,000 years.

The second oscillation is in the angle between the Earth's spin and the axis of the Earth's orbit. At the moment this angle is 23.5°, and it defines the limits to the latitudes at which the Sun can be completely overhead (the Tropics of Cancer and Capricorn, at 23.5°N and S) and the latitudes beyond which there is at least one day of the year on which the Sun never sets, or rises (the Arctic and Antarctic Circles, at 66.5°N and S). This spin angle precesses as in a gyroscope, and it varies between 21.6° and 24.5° with a period of 41,000 years. We are used to thinking of the tropics and the polar circles as immutable locations on the Earth's surface, but they are not; they actually move north and south through a range of 270 km.

Finally, the third oscillation is in the time of year when the Earth is closest to the Sun in its elliptical orbit, called the *perihelion*. This date also precesses, with a period of 23,000 years, and at present it is December. For northern hemisphere-dwellers it may seem perverse that the depth of winter is the time when the Earth is closest to the Sun.

Each of these three oscillations changes the distribution of radiation through the year and through different latitudes. The differences are slight, but they have an impact because the Earth is a lop-sided planet, having most of its land masses in the northern hemisphere and most of the oceans in the southern. The differential absorption of radiation by land and sea allows the Milankovitch radiation variations to cause changes in global climate. Therefore we would expect the average temperature of the Earth to go through a smoothly varying curve representing the sum of the three cycles, with wavelengths of tens of thousands of years. The astronomical climatic forcing due to the sum of the Milankovitch cycles does indeed vary smoothly, but the temperature response of the Earth does not. Let us see why.

THE RECORD OF THE ICE CORES

Figure 4.2 shows the climate record for the last 400,000 years or so, from the analysis of ice cores for temperature (from O^{18}–O^{16} ratios), carbon dioxide and methane levels (from air bubbles). Two results stare out at us. First, the changes are in step: warm temperatures, during an interglacial, correspond to high CO_2 (carbon dioxide) and CH_4 (methane) levels, and low temperatures correspond to low CO_2 and CH_4 levels. Secondly, the record is not a smoothly curving variation like the Milankovitch forcing, but is a sawtooth. The Earth emerges with a jerk from an ice age, warming quickly through maybe 10°C in 1,000–2,000 years to reach an interglacial state, then gradually it starts to cool off again, taking about 100,000 years of slow and fairly steady cooling to reach the depths of the next ice age. What is remarkable is that the four ice ages covered by this 400,000-year record differ in duration, but have very similar temperature and gas histories. The temperature falls off slowly and fairly linearly, except for noisy wiggles, until we are in the depths of the next ice age. The CO_2 level falls from about 280 ppm to 180 ppm, and the methane concentration from 700 ppb (parts per billion) to 400 ppb. On recovery these three quantities spring back to the values that they had in the previous interglacial. We are now in an interglacial similar to at least three previous ones. How and why does this happen?

There are many questions raised by this remarkable climate record, some of them unanswered. First, why do the CO_2 and CH_4 levels

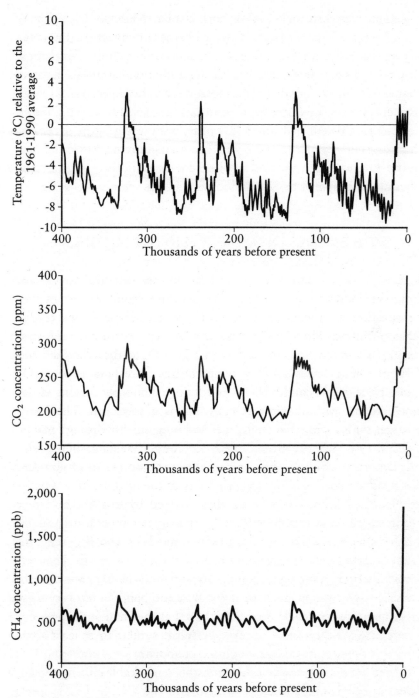

Figure 4.2: Climatic record for the last 400,000 years, from ice core evidence. Note the similarity in shape between the global temperature and greenhouse gas curves.

oscillate between such well-defined limits? When we looked at the evidence for ice ages in the early parts of the planet's history, they were one-off affairs, with various possible causes, but all of them led to one-off final states. For at least the last million years we have had a predictable history of ice ages (matching the Milankovitch forcing, except for the sawtooth nature of the response) and each one has taken us through the same cycle of temperature and the same cycle of CO_2 and CH_4 levels. On this basis, using Milankovitch forcing (and setting aside for the moment the irreversible things that we are currently doing to our planet) we can forecast how long it will be before we sink down to the next ice age, how long it will last, what the global temperature will be, and that the CO_2 and CH_4 levels at the depth of the ice age will be 180 ppm and 400 ppb respectively. We can argue from repetition that this cycle will continue. Yet of course it had a start, and this brings us to the next question. When, and why, did the cycle of astronomical-driven ice ages start? As I mentioned a moment ago, the ice core record tantalizingly takes us back only a million years, since the last few centimetres of an ice core, right down against the bedrock, are not only extremely squashed but also slightly melted by geothermal heat, so we can never get further back than a million years in this way. The most distant parts of this record are less trustworthy because of the extreme squashing to which the ice has been subjected, and it is the most recent 400,000 years, with four glaciations, that offers the best evidence of regularity in climate oscillations (fig. 4.2), although the older data seem to show this same pattern continuing.

There must have been a starting point, when the Earth's climate drifted into a state where Milankovitch cycles were just enough to drive a change from a glacial to a non-glacial world and vice versa. Before that it was presumably too warm for glaciers to form, even with a Milankovitch minimum. It looks as if this start point for modern ice ages was at the end of the Pliocene, and we are not exactly sure how many glaciations there have been. The 1-million-year ice core record contains six or seven glaciations, so there may have been up to twenty all told, although we have no way of knowing if the earlier ones conformed to this repeatable history shown by the most recent four.

The next question is, why the sawtooth? We can answer this in qualitative terms, that it is easier to destroy an ice sheet than to grow one. As the air temperature goes down because of astronomical forcing, an initially ice-free Earth starts to retain a winter's snowfall on high-altitude, high-latitude surfaces through the subsequent summer. Then next winter's snow piles on the remnants of the previous winter's snow, and permanent snowfields form that gradually grow thicker and transform themselves into glaciers and ice sheets. This is slow, but gradually causes a negative albedo feedback that multiplies the cooling effect. A glacial world slowly develops. Then comes a reversal of the astronomical forcing. The air temperature warms up. At this point the gradually formed glacier surface can melt away quite quickly, whereas it could only grow at the rate of one winter's snow layer per year. There is a physical limit on glacier growth rate but not on glacier melt rate. When the ice sheets follow a sawtooth pattern, so does the climate, with air temperature being modulated by the amount of ice present, so that air temperature too follows a sawtooth. Thus the 'sawtooth driver' is ice sheet volume, or at least surface area.

The next question is, why do CO_2 and CH_4 levels keep pace with the warming and cooling? Does one drive the other? Today we are used to the concept that the additional CO_2 that we are putting into the atmosphere warms the climate and drives a further reduction in ice sheets, among other effects. The CO_2 is the driver and climate change is the response. But during the glacial–interglacial cycles did the CO_2 rise first, for instance, and cause the ice sheet to melt, or did the ice sheets melt purely in response to the air temperature rise, with that same rise causing an increase in plant growth whose respiration causes CO_2 to rise? Experiments to test time lags between the temperature, CO_2 and CH_4 levels, to determine which drives what, come up with equivocal results: we are not yet sure what leads and what follows. And in fact the situation is more complicated than we at first imagined. If plant growth starts to increase as temperature warms (which is highly likely provided the warming is over land-based areas), then carbon from the atmosphere is trapped in the increased biomass. There is then an annual cycle which takes CO_2 from the atmosphere in spring, uses sunlight, and transforms it into sugars and lignins and other carbon-based structures, which last a long time and do not expire. So the puzzle here is that plant growth, at least initially, probably sequesters, or traps, CO_2. Another theory now gaining quite good credence is that the slight changes in temperature lead to out-gassing from the ocean surface: as the temperature increases so the ocean surfaces start to out-gas CO_2 which drives the greenhouse effect, increases the water-vapour concentration and creates a feedback cycle for further increased temperature.

Another, and extremely worrying, question is: do the glacial and interglacial temperature, carbon dioxide and methane levels represent two natural end-points for the oscillating state of the climate system? If so, there is a 'natural' climate sensitivity to CO_2 which can be calculated by dividing the rise in temperature between a glacial and an interglacial extreme by the increase in CO_2 level. Does this sensitivity tell us how, over many decades or even centuries, our climate will adjust to the indignity of receiving a large input of extra CO_2 at our hands? What if we applied the glacial–interglacial sensitivity to estimate what will happen when the climate has time to fully adjust to what we are doing to it? The result is terrifying, and I will deal with it further in a later chapter. Suffice to say for the moment that the sensitivity calculated in this way[3] is no less than 7.8°C for a CO_2 doubling, which is enough to produce a 3.6°C temperature rise just due to our present CO_2 levels. This obviously has not yet come about, but may well do so given time. This high value is known as the *Earth System Sensitivity*; it is much higher than the short-term sensitivity of the climate to added CO_2, but it indicates what will happen to our climate, eventually and perhaps over hundreds of years, if the elevated CO_2 levels are not brought down.

HOW WE EMERGED FROM THE LAST ICE AGE

The long history of ice in Earth's climate is now coming up to date, as we look at the past 12,000 years – an instant as far as Earth history is concerned – in which the planet emerged from the last Ice Age, developed a (temporarily) stable climate, and ever-resourceful mankind was able to invent agriculture and thus develop cities, architecture, money, mathematics, armies and science. Art and probably music existed back in the Ice Age; the other blessings (or curses) stem ultimately from agricultural surpluses and the need to protect fixed fields from interlopers.

The emergence did differ in one way from previous ice ages, in that we seemed to go back temporarily towards the ice age from which we had just emerged. This event was called the *Younger Dryas*, after an alpine plant called Dryas, and to distinguish it from an earlier event called (naturally) the Older Dryas. The last Ice Age reached its peak about 20,000 years ago, and then began the rapid melt representing the steep part of the sawtooth pattern. By 12,800 years ago this was bringing us close to modern temperatures, but then suddenly the temperatures in the northern hemisphere and tropics went back towards glacial temperatures for about 1,300 years before we rapidly hauled ourselves out again. There is no record of this event in Antarctic ice cores, so it was clearly a northern hemisphere phenomenon, but it took the temperature of Greenland down to 15°C below its present state. There are many theories for what caused this event. One of them is that a glacial lake, called Lake Agassiz, existed where Hudson Bay now is, and was held back from draining into the ocean by an ice sheet covering, and extending down from, Baffin Island. As part of the recovery from the Ice Age this ice sheet retreated, the damming ceased and a huge mass of fresh water drained into the Atlantic Ocean. This added a cap of fresh water which stopped convection in the Greenland and Labrador seas and so slowed down the thermohaline circulation (see Chapter 11) and brought the climate back towards cold conditions again. It is a nice and plausible story, but there is no observed evidence for it. Wally Broecker of Columbia University, who thought it up, spoke of an 'armada of icebergs' being carried into the Atlantic by this collapsing lake, which is an appealing image though probably not much more.

After the Younger Dryas interlude, the climate quickly warmed to a level slightly warmer than today, which it reached about 8,000 years ago, and from then on the climate has been extraordinarily stable. Of course we now realize that we are in an interglacial period and that this stability is illusory. In fact, from the year AD 1000 until the Industrial Revolution we were slowly cooling, as shown by the famous Mann–Bradley 'hockey stick' record of global temperatures[4] (fig. 4.3), where the long 'handle' of the hockey stick is the slow cooling and the short 'blade' is the rapid warming since the mid-nineteenth century. But the climate has been more stable, and for longer, than during any of the last four interglacials, and this gave an opportunity for *Homo sapiens*, who had survived the last Ice Age by hunting, to learn how to plant crops and stay in one place while they grew. To avoid others encroaching on the area of land where he had planted his seeds, the new farmer needed an organization that would recognize his right over that land, measuring it (so developing mathematics), writing up deeds of entitlement (so inventing writing), and protecting his rights against invaders and encroachers (so inventing police and armies). None of these inventions were needed by an Ice Age hunter-gatherer. And the man who planted these crops had several months every year with very little work to do, and so he could be organized into building monuments and megalithic temples and tombs on behalf of the organization that was providing his protection. He also had time to think about art and philosophy and, eventually, science.

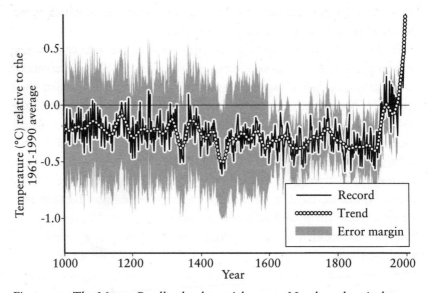

Figure 4.3: The Mann–Bradley hockey stick curve. Northern hemisphere temperatures for the past 1,000 years.

Our modern world was born when the first set of seeds, picked from the previous year's grass, was planted to provide food for the next year. All the good and all the ills. We owe the whole of human civilization to the stability of our interglacial climate.

Also, let us remember that it is not obvious that this momentous development was initiated by a 'he'. In a hunter-gatherer society (like the Inuit today), the males carry out the dangerous hunting while the females gather berries and other edible plants. It may have been the women who noticed that edible grasses sprang up at repeatable sites and could be deliberately cultivated.

After the end of the Ice Age sea levels rose rapidly, so that early coastal settlements were drowned and the record of them must be sought offshore. The land mass connecting Britain with the rest of Europe, called 'Doggerland' by archaeologists, was flooded by 4200 BC to create the North Sea and the English Channel. This process was essentially complete by 5,000 years ago, after which sea level remained remarkably constant until the twentieth century, when it started to creep upwards again. This means that the whole history of civilization has occurred with a stable sea level. We can see this around the shores of the Mediterranean,[5] where ancient coastal cities remain and where Roman fish traps, stone basins into which water was admitted to trap fish, can still be used.

In the northern seas conditions were distinctly warmer than today; it was called the 'medieval warm period'. The Vikings colonized Greenland just before AD 1000 and were able to grow hay for their animals. But from 1400 a climatic deterioration set in, sometimes referred to as the 'Little Ice Age', and ultimately the colony disappeared; climate is blamed but we do not know any details. It does appear that the Norse settlers did not want to give up their European customs and methods, and so tried to continue to keep domestic animals, failing to copy the life of the Inuit who were coming into contact with them and who depended on sealing. Hanging on to familiar habits in the face of changing conditions may have been fatal for them.

HOW LONG UNTIL THE NEXT
ICE AGE?

Estimates from projecting the Milankovitch forcing into the future suggest that we will be significantly colder in about 23,000 years' time, and will be well into the next Ice Age. But could it be that the massive additional warming that we are now inflicting on the planet will not only postpone, but may suppress, the next Ice Age? Until recently, climatologists said no, but now, seeing how rapidly warming is taking place and how many feedbacks are acting, there is an increasing belief that we may have altered the entire future of the planet as well as its short-term state. It could be that future ice ages are now abolished, or at least postponed. One study suggests that the next Milankovitch cycles will not produce a glaciation and we may not have another ice age for half a million years.[6] A more recent study,[7] which looks closely at how Milankovich forcings operate in the northern summer, suggests that a moderate continuation of carbon emission will postpone the next ice age by at least 100,000 years. One of the authors, Hans Joachim Schellnhuber, commented: 'This illustrates very clearly that we have long entered a new era, and that in the Anthropocene humanity itself has become a geological force. In fact an epoch could be ushered in which might be dubbed the Deglacial.' The term 'Anthropocene' was coined by Nobel laureate Paul Crutzen in 2000 to convey the fact that we are really in a new geological age (superseding the Holocene), where *Homo sapiens* is making a significant observable impact shaping the nature of the planet.

This all makes sense: before 2.6 million years ago the Earth was about 2–4°C warmer than today, too warm for Milankovitch ice ages. If this is the temperature required to 'turn off' (or not turn on) ice ages, then we are rapidly approaching it. It is the level to which we will return, if we remain on our present path, by 2100. Even in the Pliocene the planet needed some extra cooling to get it in a state to begin Milankovitch oscillations. So it could really be the case that what we think of as the endlessly cycling ice ages, the pattern of glacial growth, glacial flow, then melt and warming of the planet, is actually a short-lived phenomenon which depends on the planet reaching a particular mean temperature with a particular disposition of land and ocean. It could be that the set of glaciations that the Earth has had in the last 2–3 million years is all that it is now going to have, since our own impact on the climate will have taken the Earth out of a state where it can sustain periodic glaciations.

Is this a good thing or a bad thing? Instinctively I feel that any artificial messing with the climate is bad. But a fossil fuel fanatic might say that our lavish burning of the world's fossil carbon is to be applauded if it stops the next ice age, since it is only in the warm stable period since the last glaciation that Man has managed to settle down, invent agriculture, and produce the great civilizations of the past few thousand years. There could be some merit in this argument, but the problem is that we are clearly overshooting. As I will show in the next chapters, our intervention is not stopping at the possibly benign result of preventing or postponing the next ice age, but is all too likely to produce a warming faster than the Earth has ever had in its history.

5

The Greenhouse Effect

In the last chapter I looked at the natural astronomical cycle which, over periods of tens of thousands of years, takes us between glaciations, where much of the northern hemisphere is covered with ice sheets, and interglacials, where these ice sheets retreat to the fastnesses of Greenland and high mountains. Antarctica remains frozen throughout. The previous interglacial before the present one occurred about 130,000 years ago. *Homo sapiens*, who had only just evolved out of earlier *Homo* species in Africa, was not in a position to take much advantage of it. He was able to capitalize on his superior intelligence to start to spread worldwide, colonizing habitats quite different from the one where he emerged. He made stone tools and was a hunter-gatherer, but he was still too primitive, or too few in numbers, to come up with agriculture and a settled way of life. Then came another glaciation and the challenge, at higher latitudes, to just stay alive. Only in the present interglacial period has Man been able to modify his environment through technology, and only in the last 200 years has this technology involved the massive use of fossil fuels. We, the human race, are in an unprecedented situation. Our modification of the environment, of course, involves a great deal more than CO_2 production. It includes (and has done for several thousand years) clearance of land, destruction of forest, harnessing (and depletion) of water resources, and cultivation of crops. But it is our comparatively very recent invention of machines that can do work for us, and require power to do so, that has had the greatest impact. Let us see how the gases which we release cause this climatic change. But, first, let us look at the natural greenhouse effect which we depend on to exist on this planet.

THE NATURAL GREENHOUSE EFFECT

The greenhouse effect is based on very simple physics. From a single equation (based on science known by 1884) we can derive the equilibrium temperature that the Earth would adopt if it were simply a ball of matter, with no atmosphere, orbiting at its present distance from the Sun. Then we add an atmosphere and see what this does to the temperature. We will see that the natural atmosphere warms the earth – the so-called 'natural greenhouse effect' – and that the gases which we are adding to the atmosphere today warm it still more.

Let us start by imagining the Earth as a ball in space without an atmosphere and heated only by radiation coming from the Sun. As the Earth warms it radiates energy in its turn, due to its own temperature. The balance between these two tells us the temperature that the Earth reaches. Suppose this is T, measured in degrees absolute (°K), the number of degrees above absolute zero, which is the same as temperature in °C plus 273.16.

We assume that the Sun emits radiation at a constant rate. This is a function of the surface temperature of the Sun, which is about 6,000°C, and which does vary slightly over time because of the varying effect of sunspot activity, and which also has a very slow increasing trend over billions of years. But, for simplicity, we assume that at the distance of the Earth from the Sun there is a constant amount of radiation reaching the Earth per square metre at right angles to the sun's rays. This is the solar constant, S, and it is 1.37 kilowatts per square metre (kW m^{-2}). In other words, if you had a solar cell of 100 per cent efficiency on a satellite facing the sun directly, the cell would be able to generate 1.37 kW per square metre of surface, rather more than the power of a one-bar electric fire. This is the absolute maximum energy density that can ever be achieved by solar power, which is why solar energy requires such large collectors.

So we have 1.37 kW of energy in every square metre of the sun's rays reaching the Earth. How much energy in all is intercepted by the Earth? It is the solar constant S multiplied by the cross-sectional area of the Earth, which is πR^2, where R is the radius of the Earth. Some of this energy is immediately radiated back into space because the

Earth is not a perfectly black body (which would absorb all the radiation that falls on it) at visible frequencies. A fraction α is reflected straight back. This is the albedo of the Earth as a whole, and is about 0.30. So this leaves $\pi R^2 S (1 - \alpha)$ as the total radiation absorbed by the Earth.

The Earth is kept warm by this radiation but as it warms it emits radiation of its own. The Stefan–Boltzmann Law of 1884 (discovered by two brilliant Austrian physicists, Josef Stefan and his student Ludwig Boltzmann) describes how much radiation is emitted per square metre of surface by any body whose absolute temperature is T. It is found to be proportional to T^4, the fourth power of temperature, with a constant of proportionality σ, which numerically is 5.67×10^{-8} (units are watts per square metre times $°K^{-4}$). Therefore a hot body radiates vastly more than a cold body. There is another law, discovered by the German Wilhelm Wien in 1893, which describes how this radiation is distributed through the range of possible frequencies of electromagnetic radiation. At the temperature of the Sun the peak is in the visible range – so the Sun looks white, and in fact is white-hot – while cooler surfaces are only red-hot and at the low temperature of the Earth we cannot see the radiation at all, but we can measure it and find that it lies in the microwave range.

So the Earth is emitting σT^4 of local temperature radiation per square metre over its entire surface area (we assume that at these low frequencies the Earth is indeed a black body), which is $4 \pi R^2$ (the factor 4 accounts for the ratio between the whole surface of the Earth and the cross-sectional area which intercepts the Sun's rays). We can if we wish multiply the emissions by a factor ε (between 0 and 1) called the *emissivity*. At normal temperatures the Earth radiates almost as a 'black body', that is a perfect radiator, but if we retain the emissivity term we can look at the effect of adding greenhouse gases.

These two energies have to be in balance if the Earth is an isolated ball in space – the Earth emits just enough radiation to match the radiation received from the Sun, so it stays at a steady temperature T, the *equilibrium temperature* of the Earth. To find T we need to solve the following equation:

$$4 \pi R^2 \varepsilon \sigma T^4 = \pi R^2 S (1 - \alpha)$$

With slight rearranging we get:

$$T^4 = S (1 - \alpha) / 4 \sigma \varepsilon \qquad \text{(Equation 1)}$$

This is the only equation that I use in this book, but it is an important one because this simple balance between incoming and outgoing energy defines the habitability of the Earth.

The surprising solution to the equation is that T = 255, i.e. –18°C. In other words, if the Earth had no atmosphere the average temperature over its surface would be well below freezing point. We would have a frozen, dead world. Also we see that the temperature does not depend at all on the Earth's radius, just on its distance from the Sun. So our Moon's average temperature, since it has no atmosphere and is about the same distance from the Sun as we are, is also –18°C.

But clearly the Earth is warmer than –18°C. What makes it so is the fact that it is coated in an atmosphere containing gases which absorb some of the outgoing long-wave (microwave) radiation from the Earth's surface while allowing all, or nearly all, of the incoming short-wave (visible) radiation from the Sun to penetrate through. It really is just like a greenhouse, where the glass allows solar radiation to get in and warm the greenhouse, but prevents much of the long-wave radiation from getting out. So the effect of these gases is called the *greenhouse effect*.

If we look at the gases that perform this function, we see that only certain ones are important. Figure 5.1 shows radiation from the Earth as measured by a satellite passing over the Mediterranean. The smooth dashed curve indicates the theoretical distribution of energy emitted at a temperature of 7°C, from Wien's Law, which is the approximate temperature at the level in the atmosphere from which the Earth's radiation is being emitted into space. The solid curve, however, is what the satellite actually saw. The curves together show that the atmosphere below the satellite was effectively emitting energy as if it were at about 7°C, but with huge holes in different parts of the spectrum where the emitted energy is much less than that expected from Wien's Law. These are called *absorption bands* and occur because molecules can move to higher energy levels, in which an electron gains energy or the whole rotation or vibration of the molecule changes to a higher state. According to quantum theory these changes can only take place in discrete steps, where the molecule absorbs a quantum of electromagnetic energy of a particular frequency. So at certain fixed frequencies, or certain bands of frequency (as the effect is broadened out by the complexity of the molecule), the molecule will absorb some of the energy which is incident on it, allowing less energy to be transmitted onwards. If we look at energy being emitted upwards by the Earth, we find that in these bands of frequency, some of that energy is absorbed by particular gases, allowing less to pass through and out into space. This is the basis of the natural greenhouse effect. What are the effective gases? It turns out that the commonest gases in the atmosphere, oxygen and nitrogen, do not have absorption bands in the range of frequencies where Earth energy is emitted. Those that do, as fig. 5.1 shows, are water vapour (H_2O), carbon dioxide (CO_2), methane (CH_4), nitrous oxide (N_2O) and ozone (O_3). These are called the *greenhouse gases* and, although they are minor components of the atmosphere, they are of vital importance in warming the climate to the point where the Earth can contain liquid water, and hence support life.

As fig. 5.1 shows, water vapour has wide absorption bands at low and high frequencies, while methane, nitrous oxide and ozone have narrower bands at mid-frequencies. Carbon dioxide has the deepest absorption and it occurs right at the peak of the spectrum, where maximum energy is trying to get from Earth to space. From the evidence of this figure alone we can therefore expect that carbon dioxide will be the most serious greenhouse gas of all, as indeed it is.

What is the total impact of all these greenhouse gases? We see from fig. 5.1 that their effect is to reduce the amount of long-wave radiation emitted to space by the Earth. Adding up their effects across all wavelengths, it is the same as if the Earth were emitting radiation at a rate that is less than the rate that a perfect emitter, a black body, would achieve. So the effective emissivity of the Earth, ε, is less than 1 and gets smaller the more greenhouse gas is present. If we look again at our one and only equation, we can see that the left-hand side

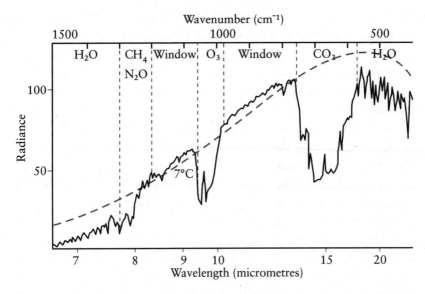

Figure 5.1: Earth radiation fluxes at the top of the atmosphere, observed from a satellite above the Mediterranean. From 8–14 micrometres the atmosphere (in absence of clouds) is mainly transparent apart from an ozone absorption band at 9.5–10 micrometres, a 'window region'. Elsewhere there are absorption bands for CO_2, water vapour, methane and nitrous oxide. Superimposed is the black body radiation curve for 7°C. Units of radiance are watts per sq m per steradian per wave number.

is being reduced, because the emitted radiation is less than the amount predicted by the Stefan–Boltzmann Law. But the right-hand side, the amount received from the Sun, stays the same. The only way for these two sides to remain in balance is for T to go up. Our equation 1 shows us that T^4 goes as $1/\varepsilon$, and so as ε goes down T goes up. The Earth has to warm up to get the same amount of radiation back out into space. This is the *natural greenhouse effect* and it is sufficient to increase T from –18°C to +15°C, the temperature that we know so well and can easily live with, the average temperature of our hospitable planet.[1]

CARBON DIOXIDE, THE VILLAINOUS MOLECULE

So far, so good. The natural greenhouse effect has made life possible. Without these atmospheric gases we would have a dead, frozen world. But what happens if we change the composition of the atmosphere? In particular, what happens if we increase the amount of carbon dioxide in the atmosphere, and so deepen the absorption in the band centred on the 15 micrometres wavelength in fig. 5.1? First of all, we reduce the emissivity of the Earth even more, so T must rise even more to maintain equilibrium. *Adding carbon dioxide to the atmosphere causes a temperature rise.* There is no way out of this conclusion. It is basic physics. To deny it is like denying gravity or asserting that the Earth is flat. Yet there are still climate change sceptics who deny any association between carbon dioxide levels and temperature. So let us put it a little more strongly: *adding carbon dioxide to the atmosphere inevitably causes a temperature rise. And, the more you add, the greater the temperature rise.* This is absolutely clear from the simple equation given above.

Man first started adding serious quantities of carbon dioxide to the atmosphere in the nineteenth century, as the needs of the Industrial Revolution outstripped water power and led to the development of coal mines, railways and the burning of coal in steam engines. It was coal-fired steam which powered the Industrial Revolution until mineral oil and electrical power came on the scene near the end of the nineteenth century (the first modern oil well, in Canada, was drilled as recently as 1858). Even then the newly developing electrical grid was supplied by power stations that were mainly coal-fired. The burning of oil became significant only with the rise of the internal combustion engine and the relentless growth in the number of road vehicles, starting with the first Benz car in 1886. As it happens, this was two years after the formulation of the Stefan–Boltzmann Law, the law which gave us the tools to recognize that carbon dioxide from combustion was warming the planet.

We could have seen what was coming for the planet, but there was some excuse for our ignorance. In retrospect, we see that the carbon dioxide level started shooting upwards in the mid-nineteenth century from its post-Ice Age level of 280 ppm up to beyond 300 (and it has now exceeded 400, nearly 50 per cent higher than pre-industrial levels) (fig. 5.2). We know this only because we can now analyse the carbon dioxide content of air bubbles in ice cores. In the nineteenth century there was no monitoring of carbon dioxide going on, and this did not start systematically until 1958 when a carbon dioxide monitoring station was built on the Mauna Loa volcano in Hawaii by the Scripps Institution of Oceanography.

Why did we miss seeing the man-induced greenhouse effect until its effects became so apparent in the late twentieth century? Well, initially there was no theory to connect global temperatures with greenhouse gas content. The vital Stefan–Boltzmann and Wien radiation laws were not discovered until 1884 and 1893, allowing the Swedish scientist Svante Arrhenius (1859–1927) to produce the first theory of the greenhouse effect and global warming in 1896.[2] So we spent the nineteenth century burning coal without the slightest idea that it could cause climatic effects.

Secondly, we had no good data on global temperatures. The UK Meteorological Office was founded in 1854 by Admiral FitzRoy (commander of HMS *Beagle* during Darwin's voyage), and similar national

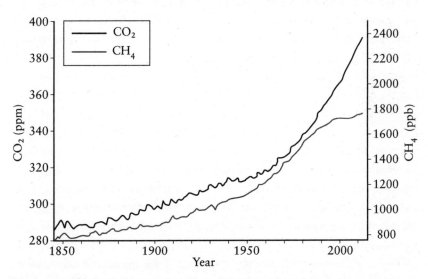

Figure 5.2: Globally averaged carbon dioxide and methane concentrations in the atmosphere (from the IPCC 5th Assessment Report).

agencies were formed soon after, but made their main business the attempt to generate weather forecasts, so it took a long time for even local climate statistics to be assembled (in England it was done by country vicars sending hand-written reports from their villages). One of the oldest continuous records has been kept, since 1767, by the Radcliffe Meteorological Station at Oxford, and has been used to show that, among other things, January 2014 was the wettest January in Oxford since records began. Of course, both for rainfall and temperature, we need worldwide data to draw conclusions based on averages, and an adequate network has been slow in developing.

Arrhenius had predecessors, but they had only been able to produce qualitative thoughts about the effect of gases on climate. Joseph Fourier (1768–1830) in France was one – we remember him now as the discoverer of the Fourier series, by which any function can be split into a set of harmonics. John Tyndall (1820–93) in the UK also had relevant ideas, and in his honour the UK has named its latest climate change research institute in the University of East Anglia the Tyndall Centre.

Arrhenius, unlike his predecessors, had the radiation laws to work with and his predictions were surprisingly accurate. He ignored the effects of clouds because he had no idea how to deal with them, but he derived our equation 1 and came up with a further, correct, analysis, deduced from this equation, about the magnitude of the warming that is produced by adding different amounts of CO_2:

> If the quantity of carbonic acid (CO_2) increases in geometric progression, the augmentation of the temperature will increase only in an arithmetic progression.

In other words, if we double the CO_2 level in the atmosphere we raise the temperature by a given amount, N degrees. If we then double it again (a much bigger increase) we raise the temperature by the *same* amount, another N degrees. This is some slight comfort in an era where our CO_2 emissions are still rapidly rising. The warming caused by CO_2 doubling is called the *climate sensitivity*. Arrhenius estimated it at 4°C; present estimates range from 2°C to 4.5°C, although, as we shall see, some estimates are much higher – as high as 7.8°C if we allow the temperature to 'catch up' with the gas level.[3] Arrhenius thought that five-sixths of the CO_2 increase would be absorbed in the ocean (it is actually about 40 per cent), and that the atmospheric concentration would increase so slowly that the doubling time would be 3,000 years. We now know that at our present rate of growth the doubling time is in fact only 75–100 years. Even with his gross underestimate of the rate of growth of CO_2, Arrhenius thought that it would have beneficial effects, ameliorating the climate at high latitudes (he did not consider ice melt) and allowing more crops to be grown to feed an increasing industrial population. He also thought that global warming would stave off the next ice age, which, as we have seen, is still a much-discussed question today.

So, finally, CO_2 was identified as the main agent in climate change, our No. 1 villain in fact, but one of its villainous properties was not recognized properly until very recently, that is, its persistence in the atmosphere. It is not that the gas, generated by the combustion of fossil fuels, remains in the atmosphere in a pure, inert state. In fact it is very reactive, and takes part in a complex set of reactions known as the *carbon cycle*. It is absorbed by green plants and oceanic phytoplankton, which in the transpiration process turn it (with the aid of chlorophyll) into biomass, emitting life-giving oxygen. This is the reaction which allows animal life to exist, so it is literally the most important chemical reaction on Earth. Having entered the structure of plants and trees, the carbon can be released again when the tree or plant dies and rots, or is cut down, or burns. CO_2 is absorbed in the ocean, but is re-emitted if temperatures and currents change. The only way to actually get rid of CO_2 from the environment is for the material produced from it to be permanently interred in the Earth's interior. The classic case is the ocean bed, where a rain of tiny shells from a common type of animal plankton called foraminifera continuously falls through the ocean after the animals die, and forms a layer of sediment. The shells are made of calcite, which is calcium carbonate, and the organism made that shell by absorbing carbon compounds through eating phytoplankton, who themselves made the biomass by taking CO_2 out of the ocean.

We still do not know for sure what the effective lifetime of carbon dioxide is in the climate system. If we release a ton of CO_2 by burning some coal or oil, for how long will that ton still influence climate, and how will its influence decay with time? The standard estimate has been 100 years, but this is a suspiciously vague number based on an incomplete understanding of the carbon cycle. Estimates of thousands of years are now being made, taking account of all the possible paths from the release of a CO_2 molecule from your car exhaust until the carbon atom in it is finally entombed in the deep ocean, or rock. But even if we take the lowest estimate of 100 years, it is clear that as we carelessly burn fossil fuel we are storing up trouble for future generations, for our children, grandchildren and great-grandchildren. They will experience continued warming due to the CO_2 which we are releasing now, just as we are still experiencing warming from factories in the First World War and gas-guzzling cars of the 1950s.

METHANE AND NITROUS OXIDE

Until recently, carbon dioxide was considered to be the main contributor to global warming, and as we have seen, there is a very clear link between fossil fuel burning, carbon dioxide increase and warming. But the other greenhouse gases also make significant contributions, which, when added together, make up about 45 per cent of the total present warming rate. One of these is methane, CH_4. Its concentration in the atmosphere shot up at the same time as CO_2 (fig. 5.2), in fact relatively more rapidly, because its present level of 1800 ppb is more than double the pre-industrial level of 700–800 ppb while the CO_2 level is only 50 per cent higher. Methane is complex and strange because it has many natural sources. These include the decomposition of organic material in wetlands (hence its common name, marsh gas) and chemical reactions arising from the actions of termites as well as digestive processes in herbivorous animals, as anyone visiting a piggery can attest. A major methane resource lies under the oceans in the form of methane hydrates (a high-pressure compound of methane and water), and of course it is the main component of natural gas. But other sources are due to Man's actions: leaks from natural gas pipelines and other aspects of coal and oil production including fracking; the cultivation of rice (because of rotting vegetation in the rice paddies); and the growth in the number of farm animals (because of an increase in meat consumption throughout the world, but especially in newly affluent countries like China). Landfill sites and waste treatment works are also methane sources. Yet despite all these anthropogenic sources, which grow as the human population grows, methane growth in the atmosphere levelled off in about 2000 and methane levels stayed steady until about 2008 when they started to rise again. We suspect that the new rise is due to offshore Arctic emissions (of which more later), but we do not understand the flattening. One possibility is that the Russians are taking more care of their gas pipelines, which previously were notorious for massive leakages; another may be that natural wetlands are being drained, dammed off and enclosed.

Despite its much lower concentration in the atmosphere than CO_2, methane makes a substantial addition to overall climate change because it is a much more powerful greenhouse gas. Per molecule, methane is 23 times as powerful as CO_2 when measured over a 100-year period; this is called its *global warming potential* (GWP). Since methane persists in the atmosphere for only about seven to ten years after emission, suffering oxidation to carbon dioxide and other chemical processes, its GWP when measured over a more limited period (the first few years after emission) is much greater than 23; figures of 100–200 have been quoted. It is clear that a sudden release of a large quantity of methane would have a huge, if short-lived, impact on climate; we consider this later in terms of what might happen from decaying offshore permafrost in the Arctic.

Nitrous oxide, N_2O, is a minor greenhouse gas, with a very low concentration in the atmosphere of 300 ppb and a long lifetime of 120 years. It originates mostly from the use of artificial fertilizers.

OZONE AND CFCS

In thinking about climatic forcing by gases we must not neglect ozone and chlorofluorocarbons, or CFCs. Ozone is the very reactive version of oxygen which has three atoms in a molecule instead of two; its formula is O_3. It is a greenhouse gas with an absorption band (see fig. 5.1) but in 1985 it became famous for another reason, the fact that an 'ozone hole' was discovered over the Antarctic using instruments mounted there by Joe Farman of the British Antarctic Survey.[4] As well as absorbing some outgoing long-wave radiation from the Earth, ozone is very good at absorbing the shortest wavelengths of incoming solar radiation. This is the ultra-violet (UV) radiation, the radiation that causes sunburn and skin cancer. It turned out that a mere 10 per cent increase in short-wave radiation can cause an increase of 19 per cent in the occurrence of melanoma, the most serious, indeed sometimes fatal, form of skin cancer.[5] Although ozone depletion due to CFCs had already been predicted by Mario Molina and Sherwood Rowland[6] (who shared a Nobel Prize with Paul Crutzen for the discovery), it was Farman's measurements that showed that over the Antarctic the ozone loss was up to 70 per cent, giving a very large increase in the UV hazard to those who dwelt within this ozone 'hole' (which extended over Australia, New Zealand, Patagonia and South Africa). The villain here was a group of chemicals artificially introduced into the environment. Chlorofluorocarbons, CFCs, are used for air conditioning and as aerosol propellants, and they react with, and destroy, ozone molecules in the atmosphere. As soon as the ozone hole was discovered, the human race for once moved quickly, and the Montreal Protocol of 1987 started the phase-out of CFCs in favour of a less harmful (but not harmless) substitute, HCFCs (hydrochlorofluorocarbons). Cynics say that this was because the substitute was readily available, but at least the ozone hole, which had spread to the northern hemisphere, is now in retreat. Despite this success against ozone depletion, HCFCs have themselves proven to be potent greenhouse gases in their own right (see fig. 5.3, which shows their significant contribution to global warming).

RADIATIVE FORCING

In order to compare the climatic impact of all these gases and other factors, scientists have developed the idea of *radiative forcing*. In our one and only equation we considered the balance of radiation on the planet Earth – the direct input from the Sun balanced by outgoing radiation due to the Earth's albedo and to its surface temperature.

Greenhouse gases act to reduce the outgoing long-wave radiation, so we can compare their power by assessing how much each of them reduces the outgoing radiation. Alternatively, one can think in terms of the added greenhouse gas being equivalent to holding the outgoing radiation constant but adding to the strength of the sun – in other words, if we compare radiative forcing with the solar constant it tells us by how much we are upsetting the natural thermal balance of our planet. If radiative forcing is counted as positive, it means that it is acting to warm our climate. Some of Man's activities cause negative radiative forcing, such as the injection of aerosols into the atmosphere, which reflect some of the incoming solar radiation.

Figure 5.3 shows the best estimate from the current – 5th – Assessment Report of the Intergovernmental Panel on Climate Change (2013). It shows a total anthropogenic radiative forcing of 2.3 watts per square metre, adding some 0.7 per cent to the average radiation received from the Sun over the whole Earth. About 55 per cent of the forcing comes from CO_2 and 45 per cent from all other sources, of which methane is the most important. This forcing has nearly

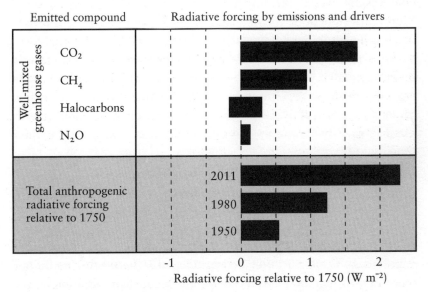

Figure 5.3: Levels of radiative forcing by different atmospheric components (from the IPCC 5th Assessment Report).

doubled since as recently as 1980 (itself a doubling from 1960), showing how rampant and uncontrolled our emissions still are, despite
decades of calls to arms by politicians, climate change activists and
some prominent scientists.

CLIMATE SENSITIVITY

In Chapter 4 we described the extraordinary similarity between the
temperature and CO_2 curves through 400,000 years of successive ice
ages and interglacial periods as shown by ice cores. This suggests
that, at least during this long period, the Earth has had two natural
levels of CO_2, about 180 ppm and 280 ppm, depending on whether it
is in a glacial or interglacial period. It has also had two equilibrium
temperature settings, characteristic of full ice age and full interglacial, and these are separated by some 6 degrees. From this we can
define a natural *climate sensitivity* for the Earth, using the temperature rise and CO_2 concentration change between glacials and
interglacials to calculate a temperature change for CO_2 doubling.
This comes out to the very high figure of 7.8°C.[7] It is tempting to
apply this figure to our modern world and question why the IPCC
gives 2–4.5°C as the present climate sensitivity when, on this basis, it
should be much higher. It must be said that one reason why this 'ice
age climate sensitivity' has not been considered very much, despite its
simplicity, is that it is just so high. If it were appropriate for modern
conditions, it would imply that only a small fraction of the potential
warming due to current CO_2 emissions has been realized so far, a
frightening thought.

Whether we take a low or high figure for the sensitivity, we can
clearly see that the Earth has not yet warmed up enough to match its
own sensitivity. The temperature rise since 1850 has been about 0.9°C
globally, while CO_2 levels have risen nearly 50 per cent which would
give 1–2.25°C of warming according to IPCC figures, and 3.9°C
according to glacial/interglacial ratios. Why the discrepancy? The
answer is the concept of *realized temperature rise*. The forcing is such
that the Earth should have warmed up much more than it has, and if
we now were able to hold the concentration of all greenhouse gases

constant, the temperature would continue to rise until it reached the figure given by the sensitivity. But at the moment it lags behind – it is constantly failing to keep up with the accelerating forcing. What slows it down? Primarily it is the slowing of global air temperature warming because of heat absorbed by the ocean, the long slow thermohaline processes (see Chapter 11) by which the deep ocean slowly turns over, gradually growing warmer but absorbing much of the extra incident radiation within its huge bulk. And 72 per cent of the surface of planet Earth is ocean surface. The good news, then, is that we are not warming as fast as we otherwise might. The bad news is that we will, inexorably, catch up with ourselves eventually – the ocean is acting as a great planetary flywheel which will ensure that warming will continue for decades to come even if by some superhuman effort we stop emitting greenhouse gases very quickly. It does make a difference what the climate sensitivity is. If it is 2–4.5°C as stated by the IPCC, then the air temperature is not lagging too far behind the forcing. But if it is 7.8°C, the air temperature has scarcely begun to respond to the forcing and there is much, much more to come, even if we cut out our emissions.

THE RECENT TEMPERATURE HISTORY OF THE EARTH

It is worth looking in more detail at just what has happened to our global temperatures since anthropogenic warming started to show itself in about 1850. Figure 5.4 shows the last 160 years of the famous Mann–Bradley curve, which has lower error bars than the previous 1,000 years because at that point we entered the era of thermometers and the beginnings of a global meteorological network. We see an interesting pattern consisting of a rapid temperature rise from the mid-nineteenth century onwards, then a pause or even a slight retreat during 1920–60, then a renewed surge. Model studies suggest that this can be explained by a rapid growth in the use of coal during the period of the pause, which launched large quantities of aerosols into the atmosphere, holding back global warming temporarily.

ARCTIC AMPLIFICATION

If we look again at the Mann–Bradley curve for the last 160 years (fig. 5.4) and compare it with a temperature curve for Arctic weather stations only, we see that they are the same shape. Figure 5.5 shows annual mean temperatures over the sea and air (SAT) for nineteen weather stations between 60°N and 90°N, and we see that the global trend of warming followed by a pause and partial retreat, then another surge of warming, is mimicked by the Arctic data. When we look at the magnitudes however, we see that, while global temperatures have risen 0.8°C between the start and end of that period, those in the Arctic rose by 2.4°C. The Arctic has warmed in a similar manner to the rest of the world but with a much greater amplitude. This is called *Arctic amplification* and in this case the amplification factor is about 3. Other estimates range from 2 to 4.

Arctic amplification is very important because it is the main reason why changes due to global warming happen in the Arctic first and why the Arctic is a bellwether for the future of the planet. Two questions which immediately arise are, what causes Arctic amplification? And has it grown in recent years?

Changes in cloud cover, increases in atmospheric water vapour, more atmospheric heat transport from lower latitudes and declining sea ice have all been suggested as contributing factors to Arctic amplification. The problem with these explanations is that, as fig. 5.5 shows, amplification has been in place since 1900, so it cannot be primarily due to a recent effect. A 2010 paper by James Screen and Ian Simmonds in *Nature*,[8] however, claims that diminishing sea ice is the driver of Arctic amplification. Their argument is that if atmospheric heat transported from lower latitudes were the major driver of warming, more warming would be expected at greater heights above sea level. On the other hand, if retreating snow and sea ice cover were the major cause, maximum warming would be expected at the surface. They showed that the heating is indeed found mainly in the lower part of the atmosphere and is correlated strongly with sea ice retreat. The problem is that the authors considered only the period from 1989 onwards, when summer sea ice was already undergoing

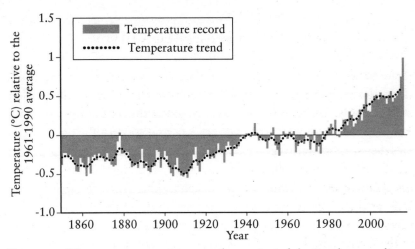

Figure 5.4: The most recent 160 years of warming of the Earth, according to the Mann–Bradley curve.

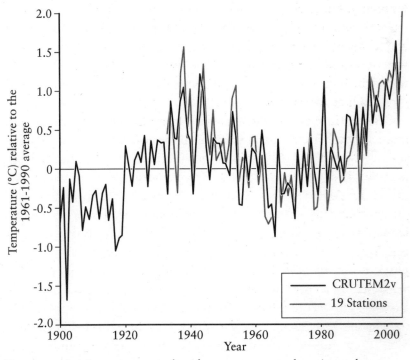

Figure 5.5: Arctic temperature data from 1900 onwards, using only measurements recorded north of 60°N. Averages from nineteen Arctic weather stations are shown with gridded regional averages (CRUTEM2v).

measurable retreat. The most we can conclude from their analysis is that the retreat of sea ice in recent years may have increased the Arctic amplification factor, but that amplification was taking place before then.

In the next chapter we show that, thanks to Arctic amplification, the extent of sea ice in the Arctic is rapidly decreasing and will almost certainly leave us with a mainly open ocean very soon.

6

Sea Ice Meltback Begins

SEA ICE IN THE NINETEENTH CENTURY

Apart from the native peoples of the north, the first travellers to keep a continuous eye on Arctic ice extent and its variation from year to year were whalers and sealers who worked in the Greenland Sea. One of the greatest of these, and the only one to combine whaling skills with scientific interests, was William Scoresby Jr (1789–1857), a native of Whitby in Yorkshire. Scoresby wrote the very first book on Arctic Ocean conditions, and particularly the variability of sea ice, in 1820,[1] and it remains a classic of polar science.

Naturally, given his occupation, Scoresby was ignored by the British establishment, although he did become a Fellow of the Royal Society. However, the Government sat up and took notice in 1818 when Scoresby announced that the ice north of Fram Strait (the strait between Spitsbergen and Greenland) was opening out and that:

> In these two seasons of 1817 and 1818, the sea was more open than on any former occasion remembered by the oldest fishermen; an extent of seas amounting to about 2000 square leagues of surface, included between the parallels of 74° and 80°N, being quite void of ice, which is usually covered by it.

This might offer an opportunity for an exploring ship to reach the North Pole, starting from the relatively high latitudes (80–81°N) that could be reached in open water east of Greenland. Scoresby himself had reached a record latitude in the earlier season of 1806:

> While I served in the capacity of chief mate, in the Resolution of
> Whitby, commanded by my Father (whose extraordinary persever-
> ance is well known to all persons in the Greenland trade,) we were
> enabled, by astonishing efforts, and with exposure to imminent haz-
> ard, to penetrate as far as latitude 81° 30'.

The Napoleonic wars had been won, the Royal Navy – easily the
largest in the world – was short of work, and great glory might accrue
from reaching the Pole or from discovering a passage to the Orient. The
Government therefore lost no time in launching a naval expedition.
They passed over the Arctic veteran Scoresby and gave the command of
the expedition to an inexperienced Royal Navy captain, David Buchan,
in the *Dorothea* and *Trent*. His second-in-command, a young lieutenant called John Franklin, was destined for a longer and ultimately
disastrous Arctic career. Their expedition to Fram Strait was a dismal
failure because they found that the drift of the ice was carrying them
southwards faster than they could work their way northwards through
it. This was a factor that Scoresby himself had not anticipated.

The Navy's exploring interests shifted to Arctic Canada and the
Northwest Passage, but Dundee whalers continued to go to the northern Greenland Sea every year, and Norwegian sealers worked in and
around the ice edge of the so-called Odden ice tongue at 75°N, an
eastward protuberance of the East Greenland ice edge where harp
seals hauled out in spring with their pups (this important little area,
vital for ocean convection, is discussed in Chapter 11). In 1872 the
Danish Meteorological Institute was founded, and for the first time
whalers, sealers and explorers had a place to send their observations.
The Institute produced a Yearbook which included month-by-month
maps of ice limits in the European Arctic, and these have since been
digitized and analysed.[2] They show no long-term trends, although
there were exceptional years, such as 1881 when a vast mass of ice
from the Arctic Ocean burst out into the northern North Atlantic
and spread almost as far as the coast of northern Norway.

The data, of course, were very crude. An entire hemispheric ice edge
was inferred from a few observations by whalers and sealers plus a lot of
experience of where the edge normally lay. This climatological approach
even lasted into the satellite era. I remember in the 1980s being shown

the creation of the daily ice chart sent out to shipping by the Danish Meteorological Institute. The only satellite available recorded in the visual range, so depended on cloud-free conditions (today we use microwave satellites which see through cloud and darkness). Cloud-free days are rare in the Greenland Sea, and if a cloudy day occurred the Institute simply reissued the previous day's chart. Their rule of thumb was not to move the ice edge until new observational data showed that it should be moved, so a week or more could pass with the ice edge shown as stationary when in fact it was simply buried under cloud.

It was not surprising that variations in sea ice extent could not be discerned easily from these data, and that sea ice was assumed to pass through a steady annual cycle, with only random fluctuations from year to year. It formed part of the great and unjustified assumption by marine scientists that everything in the ocean is constant, so that all we need to do is explore the unknown parts of the ocean more fully and add the data to great ocean atlases which, in the end, will give us a complete picture. When I first went to sea in 1969 this was still the case, and the oceanographic stations which my ship, the *Hudson*, carried out in remote parts of the Southern Ocean were added to oceanographic atlases. But, very soon after that, scientists began to suspect that the ocean is subject to enormous variability – that it has a weather as well as a climate – and the concept of world ocean atlases was gradually abandoned.

WE ENTER THE MODERN ERA

After the Second World War the Arctic ceased to be a theatre for dramatic heroic exploration and started to be a place where people needed to operate routinely. The Cold War brought air bases and long-range radar stations, and military people started to look at globes instead of Mercator maps, discovering to their horror that the shortest route for aircraft and missiles between Russia and the USA lay over the Arctic. Ice surveillance began to be carried out by military and civilian aircraft. By the early 1950s there were still no satellites, but the old sealers' observations were now supplemented by airborne surveys, since the US Navy regularly sent an aircraft out along the Eurasian ice edge ('Project Birdseye'). The Canadian Arctic was criss-crossed by aircraft of the civilian Atmospheric Environment Service, with whom I flew on ice experiments in the early 1970s in their old wartime DC-4s. In those days they were based in Gander, Newfoundland, and the aircraft had the cockpit bubbles of old Sabre fighters welded to the top of the fuselage for the ice observer to sit in and compose his chart as the plane flew along. The only entertainment in Gander was a dive called The Flyers' Club, which had a topless band, and it was here that the DC-4 pilot and crew would sit every night before an early morning take-off. Nevertheless the work got done, and the charts were of enormous value.

It was through aircraft surveys that the first evidence started to be received that the ice might be starting to retreat (fig. 6.1). The retreat was only noticeable in summer; in other seasons the sea ice filled the entire Arctic Basin up to the coasts, and it would be many more years before the autumn, winter or spring ice could measurably detach itself from the coastline.

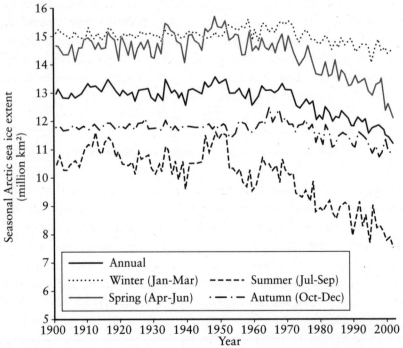

Figure 6.1: Sea ice area during the four seasons, from 1900 onwards.

By the late 1980s the retreat of summer sea ice had been recognized, tracked by the new microwave satellites,[3] and was considered to be about 3 per cent per decade, giving the ice a long lifetime. At this stage I was able to make a contribution by drawing attention to the third dimension – depth. The ice is thinning as well as shrinking. For many years I had been measuring the sea ice thickness in the Arctic Ocean from British nuclear submarines, looking up at the under-ice surface with sonar (an echo sounder) and generating a profile of draft (the immersed depth of the ice), which is 10 per cent less than thickness. I had made long voyages across the whole Arctic in 1976 aboard HMS *Sovereign* and in 1987 aboard HMS *Superb* at the invitation of the Royal Navy. In each case I organized airborne remote sensing missions by Canadian and US aircraft to supply data on the nature of the top surface of the ice to match the under-ice data.[4] When I came to compare the 1987 data with the 1976 data, both of them obtained along a similar grid of tracks stretching from Fram Strait to the North Pole, I found that there were significant differences. When averaged over the whole region covered by the submarine (see fig. 6.2), the ice in 1987 was some 15 per cent thinner than in 1976. I wrote a paper in *Nature* in 1990 on this phenomenon[5]; it provided the first evidence that sea ice thinning was accompanying the sea ice retreat detected by satellite. The progress of science here was completely dependent on submarines because no satellite technique could penetrate the ice cover and show the thickness of the ice, only the extent.

This work prompted a decade of further effort by me in the UK and by US colleagues who were analysing the more frequent American submarine missions, which usually covered a different part of the Arctic, the Beaufort Sea region. In the end we obtained the staggering result that by the 1990s the ice had thinned by 43 per cent relative to the 1970s, averaged over the Arctic and over the year. The papers by Drew Rothrock (University of Washington) and his group in 2000 on US data[6] and by me and colleagues on new British voyage data[7] agreed on this percentage despite covering different parts of the Arctic, the European sector for my data and the American sector for Rothrock's.

The implications of this discovery were of major importance, though generally unrecognized at the time by climate modellers. First, since the summer sea ice was also shrinking, this meant that the

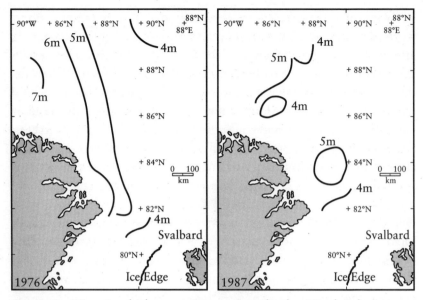

Figure 6.2: Mean ice thickness contours, Greenland to North Pole, in 1976 and 1987.

summer ice cover had lost something like 60 per cent of its volume between the 1970s and the 1990s, a far more drastic and dramatic loss than one would have suspected without taking account of the ice thickness. At this rate the summer ice would disappear fairly early in the coming twenty-first century. The world needed to be warned, and we did our best to warn it. But not only did the politicians and industrialists not want to know, neither did the scientific modellers. They continued to run unrealistic models which forecast that sea ice would remain substantially undiminished right up to the end of the twenty-first century. The UK Meteorological Office still clings to these impossible predictions. Nature would soon prove them wrong.

THE LAST TEN YEARS OF ICE COLLAPSE

The sea ice cover in the Arctic runs through an annual cycle (Plate 17) in which the greatest extent is achieved by February and the minimum extent in mid-September. It lags two to three months behind the cycle of solar radiation because it takes time for the solar radiation to do its work in melting ice and warming the sea and land. In the last decade attention has become focused on the September minimum because in 2005 this was subject to a big retreat relative to earlier years, and for the first time the summer ice cover was fully detached from the land masses of Siberia and Alaska, though it still clung to the coasts of Greenland and the Canadian Archipelago (Plate 13). The Northern Sea Route (Russia's name for the Northeast Passage) was completely clear of ice, although the Canadian Northwest Passage was still substantially blocked. The total ice area in September 2005 was only 5.3 million km^2, compared to 8 million km^2 in the 1970s and 1980s when the 'seasonal mean' (shown in pink in Plate 13) was established. I had carried out a new submarine voyage in 2004 and saw a continuing thinning, so recognized that the accelerated ice retreat was simply the ice cover's response to the thinning produced by extra warming, and that this was the start of a collapse.

In 2007, after a partial recovery in 2006, came an even bigger leap downwards in area (see Plate 13). This time a huge bite was taken out of the ice north of Alaska and eastern Siberia, to create a substantial blue ocean where ice always used to be. The ice area shrank to 4.1 million km^2. Strangely, the ice distributed itself so that this time the Northwest Passage was completely clear while the Northern Sea Route was blocked at the Vilkitsky Strait north of Siberia.

The simplest explanation of this is a further thinning and collapse based on ice melt. But a dynamic factor came into play as well. The winds during the early summer over Alaska had been from the south and west and blew the surviving ice across the Beaufort Sea and towards Fram Strait. This was shown by the tracks of drifting buoys of the International Arctic Buoy Programme (IABP). IABP is a genuine international collaboration whereby every year an array of drifting buoys is dropped over the Arctic and transmit their positions to tracking satellites. In 2007 the ice moved rapidly east, with some of the buoys overtaken by the retreat so that they fell into the sea, and the surviving ice cramming itself into Fram Strait. The exit from the Arctic Ocean was like the crush around cinema doors when someone shouts 'Fire!'

This new pattern of motion became more prevalent through the

2000s. Instead of driving a big rotating gyre, as described in Chapter 2, the new prevailing wind pattern regularly causes ice to exit through Fram Strait without achieving a circuit of the ocean. The older, multi-year ice, has been taking part in this new overall motion, drifting out of the Arctic through Fram Strait instead of circulating around in what used to be the Beaufort Gyre. So each year the fraction of multi-year ice in the Arctic has been less than in the previous year, and so it has continued to go on until there is almost none left. New ice has been simply forming, drifting to Fram Strait, and exiting the Arctic Ocean, leaving little to become multi-year ice. This dramatic decline of multi-year ice during the 2000s has been followed using a microwave satellite which discriminates between first-year and multi-year ice.[8] The dominance of young ice in the Arctic is itself a factor making the average ice thickness less, though the climate-induced reduced growth rate is of greater importance.

A PERSONAL INTERLUDE, 2007

My Arctic experience in that critical year of 2007 was quite extreme. I made a new trans-Arctic voyage in March in the submarine HMS *Tireless* to measure ice thickness, but this time using a multibeam sonar which gives a wonderful three-dimensional view of the ice underside (Plate 10), the first time that this had been used on a large scale under ice. We crossed the whole Arctic, sailing from Faslane in Scotland, entering Fram Strait and sailing round the north of Greenland and into the Beaufort Sea. In the Beaufort Sea we found that almost all the ice above us was first-year. We spent a few days criss-crossing an area under an ice camp set up by the University of Washington, called APLIS (Applied Physics Laboratory Ice Station), where a group of scientists was busy measuring the top surface of the ice by drilling holes, using electromagnetic methods to sound through the ice, and doing surveys by aircraft equipped with lasers to measure the shape of the top surface. Very important collaborative data were being generated by the submarine and camp working together when suddenly, on 20 March, disaster occurred. I was on watch with my sonar in the early evening when there was an enormous BANG, unbelievably loud, accompanied by a shock wave and a mass of brown smoke, which blasted at enormous speed along the corridor of the deck below then billowed up the stairs into the control room, accompanied by the Captain rushing up the stairs and shouting orders: 'Emergency stations! Don EBS [emergency breathing system] masks!'

Everyone was frozen in terror. Then everyone rushed for the nearest masks. I had moved aft to get out of the way and was beckoned into the wireless room, where the wireless operator gave me a mask and plugged it into the oxygen line. He was really worried. 'This is serious. I've never seen anything like this before!'

Nobody knew what had happened. A collision? A nuclear accident? I expected to die within a few seconds – explosions inside submarines usually mean the end of the boat – and we all put on oxygen masks and waited for our final moments. And yet I felt absolutely calm. I was so far beyond terror that I did not panic at all. I just put on my mask, sat down in the compartment and awaited death. I don't even think that my heart rate went up. This is the closest that I have ever been to death in the Arctic and, curiously, it did not bother me. But the lights stayed on and the boat kept moving.

Each compartment reported, until it was found that the explosion had happened in the forward escape compartment, and that what had blown up was a SCOG (self-contained oxygen generator), known to submariners as a 'candle'. A submarine supplies itself with oxygen for the crew to breathe by electrolyzing water, but if a boat's electrolyzers fail, for example through freezing up, it must renew its oxygen in a different way, and it does so by introducing a canister of potassium chlorate into an apparatus within which a catalytic reaction takes place that releases oxygen. One of these canisters had exploded when it was being loaded into the SCOG and the consequences were dire. The boat was filled with toxic gases (carbon monoxide to high concentration, as well as carbon dioxide) and smoke.

Things then got worse with the cry of 'Fire! Fire!' This is the greatest fear of submariners – fire at sea, and even worse, fire under the ice, with no way of getting up. The fire was put out, partly by flooding the affected section and partly by firefighting by the crew. Twice the fire restarted, with more cries of 'Fire!' before it was finally put out.

The immediate need was to surface. By enormous good fortune we were close to a polynya. We had been plotting each polynya as we passed it and the last one was close. We turned, sailed straight into it, stopped, checked for ice above each of the upward-looking sonar transducers (bow, fin and stern) and as soon as they were clear, came straight up. Thanks to the skill of the Captain, we were almost wholly within the polynya. After more manoeuvring, and then a tense wait as we rose towards the surface, the blessed words came over the PA: 'We are surfaced in a polynya. Stand by to open up.' The hatch was opened and ventilators blasted fresh air through the boat to clear away the poisonous fumes.

Meanwhile there was talk of a casualty or casualties, who were being taken to the Junior Rates Mess (the mess for junior seamen, which can be converted into a sick bay) for treatment by the Medical Officer. It didn't sound too bad at first. But then the terrible rumour went round the boat and reached us: 'There's two dead!'

A young sailor burst into the radio room, crying, and sobbed out the story to the radio operator. He had seen the bodies. Only later did I learn the full, terrible truth. There were indeed two dead, two sailors, one aged eighteen, one aged thirty-two. The older sailor had been celebrating his engagement only the day before. The two sailors had been sent to let off a 'candle'. It exploded as they did so, and the metal casing of the SCOG apparatus acted like shrapnel, killing them. Bits of the apparatus were embedded in the deckhead of the compartment, and the plates of the deck were twisted by the blast. The sailors didn't stand a chance. And their bodies blocked the hatch so that it was hard for the firefighters to get in. A third casualty wasn't so bad – he had inhaled a lot of smoke in the initial burst, but was reasonably OK.

We had already sent an emergency message to APLIS, and soon after we surfaced a group of Americans arrived by snowmobile in the darkness from the camp, bringing more medical supplies. The walking casualty was evacuated and flown in moonless darkness by helicopter straight to Prudhoe Bay, where a C-130 was waiting to take him to Elmendorf Air Force Base in southern Alaska for treatment. The two bodies were carried out and taken to the camp.

The whole safe, ritualized world which I had become used to in six submarine voyages had dissolved away, leaving horror and terror in its place. Every irrational fear that I had entertained in past voyages had now been realized, and yet now that the worst had happened I found myself unafraid and still completely calm. My colleague Nick Hughes who was with me felt the same. We spent the night on board and were taken off in the morning. Many times, jokingly, the officers of the submarines had told me how safe I was – I was in an even healthier environment than on the surface, because the submarine reactor was well shielded while our depth underwater meant that we were exposed to lower doses of cosmic rays than the supposed unfortunates who live on the surface of the Earth. Life in a submarine seems misleadingly safe – you work in shirtsleeves, eat excellent food, and sit on chintz-covered chairs in a comfortable wardroom. Yet my fifty Arctic field trips using highly uncomfortable tents, huts, ships, planes, helicopters, dog sledges and skidoos never brought me closer to death than this one episode in a submarine.

The end of the story is not very elevating. There had been standing orders to check SCOG canisters for cracks, as just such a fire had started in a SCOG on board the Russian space station Mir, caused by oil leaking into a crack in the canister to create an explosive mixture. Dutifully the ship's crew had sent several defective canisters back, only to find that the naval base had put them back on the submarine and ordered the crew to use them in order to save money. There was a Board of Inquiry afterwards which reported on 12 June 2008. There has not been another British submarine voyage to the Arctic since.

Amazingly, given the circumstances, I flew back to the UK for a week then flew back out to the APLIS camp to do another experiment at the site of this disaster, using an AUV (autonomous underwater vehicle) to map a small number of pressure ridges in great detail. One of these (Plate 8) was a first-year ridge which had formed only seven days earlier, so I could see the shape of a newly formed ridge where the ice blocks are simply piled up loosely like a linear slag heap, with little or no strength. This kind of ridge is now prevalent in the Arctic, and the old massive multi-year ridges, which were a huge impediment to icebreakers, have now almost disappeared. The AUV study was good therapy: physically, I had a serious cough for many months afterwards, but mentally I seemed to be fine.

THE NEXT STEP DOWNWARDS FOR THE ICE — 2012

Although the trend of Arctic sea ice extent has been relentlessly downward, and is accelerating, it is not uniform. Random weather factors can intervene either to accelerate or retard the summer sea ice retreat. If a random factor allows a partial recovery in a given year, this is always hailed by climate change sceptics as a sign that Arctic sea ice is not retreating at all. The subsequent enhanced retreat in the following year is ignored.

The graph of ice extent in September (Plate 13) demonstrates these large fluctuations, but also shows the powerful trend (see also Plate 17). After the dramatic year of 2007 the ice extent hovered slightly above the 2007 minimum, until in the summer of 2012 a new record low occurred, of only 3.4 million square kilometres. This time the loss was distributed around all longitudes, a real circumpolar withdrawal of the ice rather than a bite caused by a dominant wind direction. In this case the achievement of a record low area was assisted by a storm later called the Great Arctic Cyclone,[9] which hit the Arctic on 6 August. It was the most intense summer storm since satellite monitoring of climate began in 1979. The sea ice was already approaching its summer minimum, and, according to a satellite study by Claire Parkinson and Joey Comiso of NASA Goddard Space Flight Center,[10] the storm then caused an area of 400,000 km^2 of ice to separate from the main pack, to break up under wind and wave action, and ultimately to melt. A different modelling study by Jinlun Zhang and colleagues[11] (University of Washington) suggested that a more modest 150,000 km^2 was lost as a direct effect of the storm, but both studies agreed that the storm had a measurable impact at a critical time.

THE FINAL YEARS OF SUMMER ICE

In 2013 there was less storm activity during summer, and the wind directions during the storms tended to move cold air into the Arctic, depositing fresh snow on the ice which slowed its melt and raised its albedo. The year 2014 was also one in which the ice hovered around previous values, and even staged a slight recovery. Its vulnerable state was very clear, however. I was out during August in the southern Beaufort Sea ice edge region aboard the US Coastguard icebreaker *Healy* and found the ice cover extremely rotted and close to total melt (Plate 2). The relentless downward trend shows that these partial recoveries, or rather wiggles on the way down, are characteristic of the ice decline and we expected another downward lurch in 2015, especially as 2015 was a year in which there was a partial 'El Niño', a change in the pattern of winds and currents in the Pacific Ocean which has the net effect of releasing stored oceanic heat and warming the atmosphere faster.

In fact 2015 gave us the fourth lowest area of ice in September ever (Plates 13, 17), and El Niño continued with greater intensity so that 2016 remained a possible candidate for an ice-free September. Modellers still pretend that an ice-free summer will not happen until 2050 to 2080, but the observational data show that this is quite impossible. One modeller who does predict a rapid demise for the summer ice, soon after 2016, is Wieslaw Maslowski of the Naval Postgraduate School, Monterey,[12] who has two advantages: his model can represent very small-scale processes (it has a 2.4-km grid scale, the finest scale of any climate model), and he has the use of one of the world's most powerful computers, that owned by the US Navy in Monterey. He also emphasizes processes which are ignored or treated crudely in other models, notably the role of heat in the upper ocean in melting ice and in changing the state of the so-called mixed layer, the shallow zone just under the ice–water interface.

The importance of the year-to-year random factors can be seen by looking at a map of ice concentration at the time of maximum retreat. Plate 14 shows a satellite map of ice extent on 20 September 2012, the day of maximum retreat. The map was produced by the University of Bremen using a different technique from the more familiar maps from NSIDC (the National Snow and Ice Data Center, Boulder), one which shows ice concentration within the ice limits rather than just a white expanse. We can see from the map that moving in from the ice edge in the Beaufort Sea and the Russian Arctic seas there was a wide fringe of low-concentration ice, such that two or three more days of melt, which could easily have occurred through random weather factors, would have removed another large area, creating a record even lower than the 3.4 million square kilometres that actually occurred.

WAVES IN THE OPEN WATER

One of the random weather factors involved is waves, and this was certainly true of the big storm of 2012. Waves are bound to become even more important as the summer sea ice area shrinks further, but even a cursory look at the 2007, 2012 and 2015 maps shows us that huge areas of open water now exist around the fringes of the ice in summer. The enormous retreat of sea ice is creating enough open water area to permit winds to generate plenty of wave energy at the ice edge in previously sheltered seas such as the Beaufort. This may be enough to break up the ice and speed its melt and retreat even further. In other words, warming causes sea ice retreat, which opens up a large stretch of open water, which permits wave growth, which interacts with the ice to cause break-up and decay, which opens up the ice still further. This is the first of the great Arctic sea ice feedbacks, of which I will discuss more in Chapter 8.

The study of waves and their impact on ice is relatively recent, and I have been involved since the early stages. In fact it was the topic of my PhD thesis in 1973. I joined the Scott Polar Research Institute in Cambridge in 1970 as a research student. The Director, Dr Gordon Robin, had worked both on glaciers and on sea ice, and agreed to take me for a sea ice project. The project that I settled on was to understand what happens when ocean waves enter sea ice. At the time, very little was known about this. In those days few oceanographers worked on polar problems, so anyone entering the field had a vast range of unexplained phenomena to choose from. Dr Robin had made a voyage to the Antarctic on a ship with a wave recorder aboard, measuring wave energy at different distances inside the ice edge, and had sent a research assistant out to do the same. But that was all – two data sets and no theory of what was going on.

Fortunately a second, healthy, aspect of those days was that computer modelling had scarcely been invented and the emphasis was on solving scientific problems by getting out to make measurements in the field. I wish that were still the case today. Gordon Robin was assiduous in finding field opportunities for me. In February 1971, only four months after I arrived, he made use of contacts in the Navy (he had served on submarines during the War) to send me out to study waves in ice from the diesel-electric submarine HMS *Oracle*, which was sailing to the ice edge in the Greenland Sea to escort the first British nuclear submarine going into the Arctic Ocean, HMS *Dreadnought*.

I had a wonderful time in *Oracle*. It was exceptionally cramped, dirty and smelly, but fascinating. It was all one tube, with no separate decks – the control room, diesel engines, battery compartment and torpedo tubes all led out of one another, just as in Second World War U-boat films. Again, just as in the War, the crew wore grease-stained white wool sweaters, never washed, and slept in bunks which were crammed into spare corners all over the boat. My own bunk was down at deck level outside the wardroom door, and there was another bunk a couple of inches above my nose. This was slept in by the wardroom steward and turned into a table in early morning, where he dished out breakfast materials from the galley on to plates. When we were on the surface and rolling, the breakfasts would often spill and run down the side of my bunk. The inspiring Captain, Hugo White (later Admiral Sir Hugo White, Commander-in-Chief Fleet) dived us under the ice edge and surfaced through ice several tens of kilometres inside the ice pack, a bold manoeuvre for a diesel submarine which needs to recharge batteries. The Captain presented long-service medals to crew members in a ceremony on an ice floe.

For my scientific work, I carried out a procedure that had been proposed several years earlier by my scientific hero, Walter Munk, of Scripps Institution of Oceanography, to hover under the ice at different distances inside the ice edge and use the submarine's upward-looking echo sounder to measure the range to the sea surface and hence record a time series of waves, since the submarine itself was deep enough not to be affected by wave action and so acted as a stable platform.[13] I got some excellent data on wave decay in the ice,[14] which were the first really accurate field measurements and which showed that the decay with distance took an exponential form. This demonstrated that waves were being reflected by the ice floes, with the reflected energy going off in all directions and reducing the strength of the waves that were still trying to penetrate further through the ice. I developed a theory for this process, which is called *scattering*, did more field work from an aircraft laser, and wrote my thesis in 1973.

A few years later, back in Cambridge again after a period in Canada and a year as a visiting professor at the US Naval Postgraduate School in Monterey, I received a grant from the US Office of Naval Research (ONR) to study wave decay in ice intensively in a project called MIZEX, the Marginal Ice Zone Experiment.[15] Much later, in fact in 2012, ONR renewed its interest in waves in ice when, like many ice scientists, they suspected that the waves generated in the summer open water may indeed be enough to tip the balance towards the destruction of the summer ice. I have now joined a large group of scientific partners to study the wave–ice phenomenon using modern methods. We are using satellite-tracked wave buoys to record wave energy in the ice. Some drift over large areas of the Arctic Ocean, while others were launched only for short periods in ice edge regions during an October–November 2015 voyage in the new University of Alaska icebreaker *Sikuliaq*.

Our experience on the *Sikuliaq* demonstrated another climatic aspect of wave–ice interaction. We found that the early autumn refreezing, when the ice edge normally advances rapidly from the southern Beaufort Sea down into Bering Strait and into the Bering Sea, was just not happening as the textbooks predict. Instead the ice edge would advance, with new ice forming initially as pancake ice because of the waves (more about this ice type in Chapter 11), but then a real storm would arise and the new ice would vanish, melted away by the heat in the water column that was brought to the surface by the waves. This heat had accumulated in the water during the ice-free summer. The battle between the advancing ice and the resisting sea was bound to be won by the ice in the end, but the sea fought a delaying action, and the result as I write (May 2016) is that the ice extent is at a record minimum for the time of year, suggesting a very low ice extent to follow in September 2016. It is also significant that globally February 2016 was the warmest seasonally adjusted month

1. Canadian Scientific Ship *Hudson* off the north coast of Alaska in August 1970, surrounded by multi-year ice (seen from the ship's helicopter).

2. Typical rotted ice in the southern Beaufort Sea in August 2014, observed during an expedition aboard USCGS *Healy*.

3. Winter leads steaming with frost smoke, Greenland Sea.

4. A typical winter landscape of smooth first-year ice covered with snow. The ice thickness is between 1 m and 1.5 m. To the right is a refrozen lead where the ice has grown to resemble the rest of the ice sheet in appearance and thickness. This scene is now typical of the Arctic.

5 & 6. (Top) An expedition camped on the Yermak Plateau, north of Svalbard in the Arctic Ocean, in the winter of 2003. A crack appeared in the ice the following morning. (Below) Over the next couple of hours a lead rapidly opened and widened, the scale shown by the same tents.

7. One-week-old pressure ridge in the Beaufort Sea, April 2007.

8. The same pressure ridge in the Beaufort Sea mapped by multibeam sonar on a small AUV. Colour scale is draft in metres. The area in the red circle was visited by the diver (*see inset*).

9. A stamukha found drifting in the Greenland Sea, July 2012. The yellow base station visible on the summit was used to produce a topographic scan of the ridge.

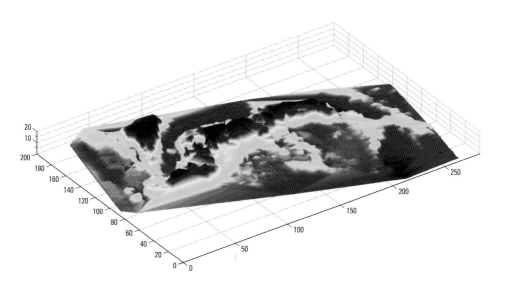

10. A multi-year pressure ridge mapped by the multibeam sonar aboard HMS *Tireless*, March 2007. Distances and heights are in metres.

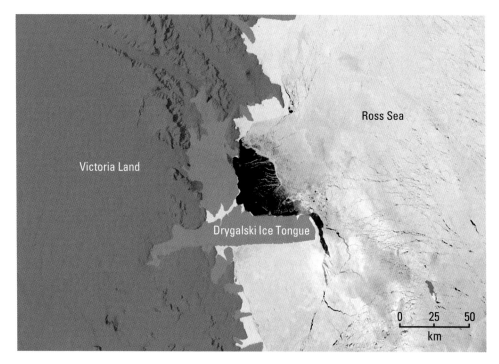

11. The Terra Nova Bay polynya in the Ross Sea, October 2014. The open water of the polynya is dark, covered by the white streaks of clouds driven by katabatic winds off the nearby ice shelf.

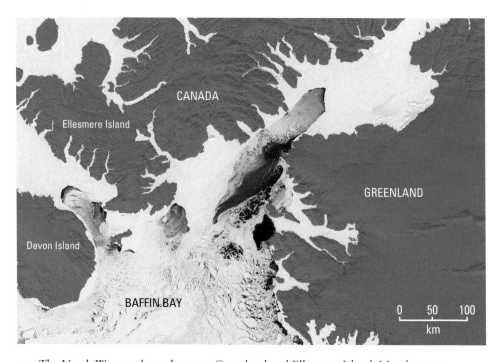

12. The North Water polynya between Greenland and Ellesmere Island, March 2015.

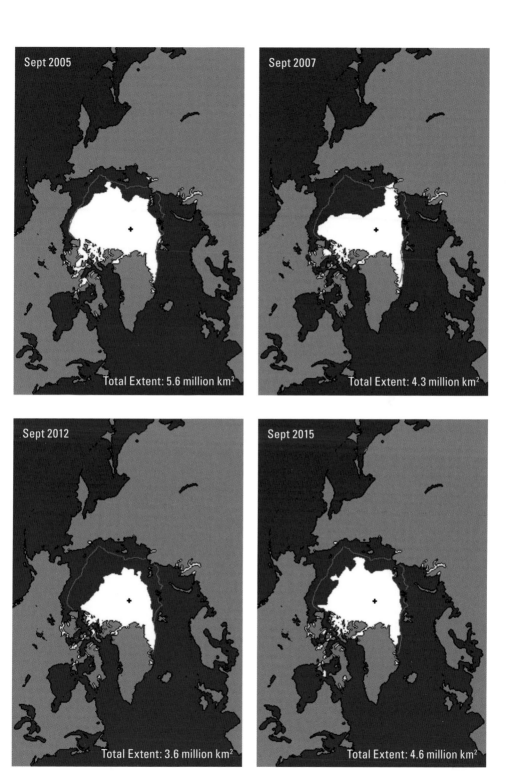

13. Sea ice extent in September of 2005, 2007, 2012 and 2015. The pink line is the long-term (past) median ice extent in September.

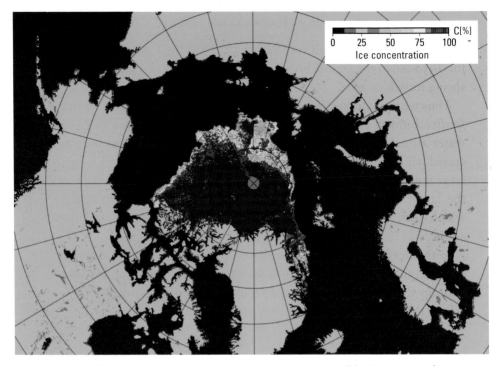

14. Mid-September 2012 ice extent and concentration as retrieved by University of Bremen, showing very low ice concentrations near the edge.

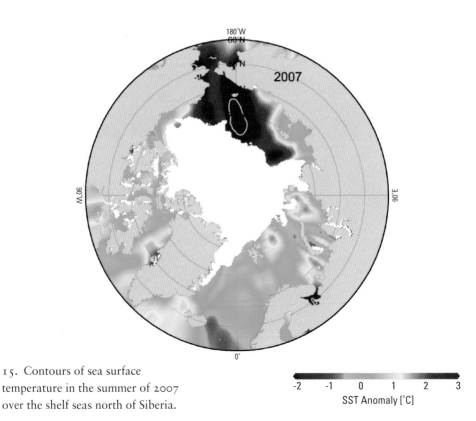

15. Contours of sea surface temperature in the summer of 2007 over the shelf seas north of Siberia.

since records began, a full 1.35°C warmer than the February average for 1950–80. This record-breaking then went on to be true of every spring month of 2016.

Thus we have discovered, just in the past few months, that *wave-ice feedback* takes two forms. In high summer the vast area of open water around the edges of the Arctic Ocean permits waves to be generated which penetrate into, and are scattered by, the ice, breaking up large floes into smaller fragments and speeding their melt. In autumn the larger storms cause wave-induced mixing of the upper waters, which brings up heat absorbed during the summer and holds back the advance of the ice by melting the new ice as it tries to form.

In the next chapter we follow the decline of the ice to its ultimate conclusion, an end to the summer sea ice cover. Then in Chapter 8 we take up the problem of feedbacks again and describe other serious effects that the ice retreat is having on global processes.

7

The Future of Arctic Sea Ice – The Death Spiral

WHAT NEXT FOR SEA ICE?

A clever and modest geophysical data analyst, Andy Lee Robinson, came up with a way of representing the Arctic ice volume data which makes it very clear how rapidly the decline in summer extent will lead to disappearance, and how the other months of the year will follow the downward trend. We begin with fig. 7.1, which shows the volume anomaly (that is, ice volume compared with a 1976–2015 average) plotted against time. This builds on the decline in extent shown in fig. 6.1 by making use of thickness data as well. When we multiply thickness by area to get volume, the relative rate of decline increases, since both area and thickness are shrinking together. The graph shows a linear best fit to the trend, but the data from 2002 onwards indicate an acceleration in the rate of decline. Because of the thinning, the ice is likely to vanish more quickly than if we looked at the extent data alone.

The accuracy of the two types of data that have been merged in fig. 7.1 is not identical. The ice area is known very accurately – it is taken from satellite imagery which gives us both ice extent (the area within the outer ice edge) and area (the actual area covered by sea ice, allowing for open water within the pack). The thickness used in fig. 7.1 is not so well known, and is derived from a simple model, since thickness has not been monitored over the whole of the Arctic. A satellite now exists which is designed to carry out this task. It is called CryoSat-2 (CryoSat-1 failed soon after launch), launched in 2010 by the European Space Agency, and it uses a radar altimeter to measure the elevation of the ice surface above the waterline, called the

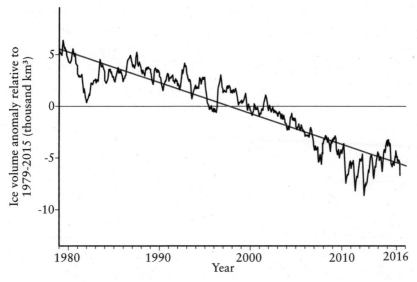

Figure 7.1: The decline of Arctic ice volume over the past thirty years.

freeboard. A radar beam bounces back from the ice surface and an accurate measure of the range tells us the freeboard. A conversion factor based on what we know of the density of ice and snow can then be applied to convert freeboard to thickness. This factor varies with time of year, location in the Arctic and type of ice, so not surprisingly there have been arguments about which factor to use. Ice thickness values obtained from CryoSat-2 have been published,[1] but only from 2012 on, and have been subject to much criticism. I am interested in the whole trend of what has happened since 1979, so I prefer to use the partial data obtained from the sum total of submarine transects under the Arctic, carried out in the US mainly by Drew Rothrock of the University of Washington and his colleague Mark Wensnahan[2] and in the UK by me. A project called PIOMAS at the University of Washington (Pan-Arctic Ice Ocean Modeling and Assimilation System) takes the thickness data and applies them to a very simple model in which the mean ice thickness is calculated for the whole Arctic Ocean based on the partial data from submarine voyages interpolated by what we know about the distribution of ice types, ages and driving forces such as air temperature. So PIOMAS is not pure data, but is the nearest thing to it that we can achieve, with minimal modelling intervention in the data analysis process. In this respect it is diametrically opposite to the ice–ocean climate models used by the IPCC.

Andy Lee Robinson cleverly visualized these results. He took the monthly data that had been used to generate fig. 7.1 and wrapped the data into a clock-shaped presentation, in which we start with 1979 at 12 o'clock and run round the dial clockwise to the present at just before 12 again. The distance from the centre is the mean ice volume in a given month of that year. The result is a set of twelve curves (Plate 16), which would be concentric circles if nothing were happening to the Arctic ice. But the curves are all spiralling in towards the centre, and the curve for September has very nearly got there. When he saw this, the glaciologist Mark Serreze, head of NSIDC (National Snow and Ice Data Center) at Boulder, called the curve the 'Arctic Death Spiral'.

If we look at the Arctic Death Spiral it is clear, in fact it is blindingly obvious, that the summer Arctic sea ice does not have long to live. The downward trend brings the summer months to zero ice cover in 2016 for an ice-free September and October, 2017 for an ice-free August to October, and 2018 for an ice-free July to November. Thus the trend lines predict two ice-free months in 2016, three months in 2017 and five months in 2018. The curves for the other seven months of the year lag some way behind this, but they are all heading downwards at an accelerating rate, all heading for the centre of the death spiral. We should not use this simple extrapolation technique to make predictions about the winter months, for much may happen in the next few decades to change the state of the Arctic during the winter. But we are perfectly justified in using extrapolation to obtain the dates of summer disappearance shown above, because *we don't have far to go*. Some new process can always arise that will modify the rate of decline, but there is no sign of such a process at the moment, and the decline only has to carry on for a couple more years at most before the September extent truly is zero. It may even have happened before this book is published.

We have seen the 'wiggle factor' in the data of Chapter 6, that a clear and strong long-term trend can, in any given year, be disrupted by random factors associated with weather events during critical phases of the sea ice growth and decay process. But the wiggles are temporary while the trend is inexorable. There is no doubt that the time series of fig. 7.1 represents a powerful trend that will lead the September sea ice to disappear very soon. The trend points to 2016, but of course the wiggles may make the date a later one — but not much later.

By the term 'disappear', scientists mean that the main bulk of the ice cover will go and that the Arctic Ocean will be open from America to Eurasia. Clearly pockets of ice will remain, especially along the coasts and in channels such as the Northwest Passage, amounting to a million square kilometres or so. But the main body of ice will disappear. As we shall see in the next chapter, every Arctic feedback that we can detect is positive and there is no process that we can think of that will slow or halt the decline of the summer sea ice towards oblivion.

Let's remember what has contributed to this recent accelerated decline. The multi-year ice has almost all gone, and even if the Arctic atmospheric circulation suddenly changes, there is not time within the next year or two for ice freshly retained in the Arctic to reach substantially greater thicknesses. The warming of the ice-free ocean in summer continues unabated, and will continue to delay the autumn freeze-up still further, enhancing the rate of the break-up of existing ice by warm water melt and wave action.

HAVE WE PASSED A TIPPING POINT?

In recent years the concept of 'tipping points' has become popular, even in fields unrelated to climate, and it has become a very loose term. I am going to adopt a rigorous definition and say that a tipping point occurs when a system which has been stressed beyond a certain level does not return to its original state when that stress is removed, but migrates to a new state. Many of us came across Hooke's Law at school. A wire or a spring is stretched by a weight and the stretching is proportional to the weight applied; it springs back to its original length as soon as the weight is removed. But if the weight is too great, it exceeds what is called the elastic limit of the wire, which keeps stretching more and more with the same weight. If the weight is removed the wire does not go back to its original length, and never will because the crystalline structure of the metal has been changed. It has passed a tipping point. Has Arctic sea ice reached a tipping point? I believe that it has, for the following reason.

We know that the area of multi-year ice in the Arctic during the winter is diminishing year on year.[3] This is partly an effect of the atmospheric pressure field, which now drives ice out of the Arctic Basin by direct paths from the formation regions, instead of allowing ice to make long-period rotations in the large Beaufort Gyre. If this field continues to prevail, a greater area of sea ice will melt completely year on year, since first-year ice grows more slowly than in the past, melts more rapidly and permits a greater area of warming water to be ice-free every year. Once the ice cover has completely gone in a given summer, the following winter's ice will be all first-year, and will then melt again in the following summer. So there is no chance for a substantial multi-year ice cover to re-form. The tipping point for sea ice therefore occurs when the summer melt rate versus winter growth rate becomes such that all the first-year ice melts during the summer. Then, no first-year ice survives to become multi-year ice in October (when freezing starts), and the multi-year fraction in the Arctic cannot increase but must continue to decrease until there is none left. Then the Arctic, for ever afterwards (or at least until the climate becomes cooler again) will only have a seasonal ice cover.

Much public attention was gained by a paper by Steffen Tietsche and colleagues in 2011 which drew different conclusions,[4] but the arguments in their article are wholly misleading. The authors carried out the artificial procedure of removing the entire Arctic ice cover (in a model) and observed that within two years the ice cover recovered to its former level. This was repeated at twenty-year intervals for an ice cover declining in area in response to the modelled amount of warming, and in each case the ice cover recovered to its former state. The authors concluded that ice retreat is reversible, and that all we have to do to regenerate the Arctic ice cover once it has retreated is to reduce carbon emissions such that the radiative forcing is no longer at work. This is an unjustifiable conclusion for two reasons. First, the act of total removal in a computer model is an artificial change imposed on the ice cover without changing anything else, so naturally the ice will tend afterwards to revert to its previous state. Secondly, the conclusion that natural ice loss is reversible fails to take account of the well-known time lags involved in CO_2-induced radiative forcing, whereby a given quantity of CO_2 released into the atmosphere continues to have an impact upon the climate system for more than 100 years. Even a precipitate reduction in CO_2 emission would not cause air temperatures, for instance, to drop for many years or even decades, let alone sea temperatures.

HOW DO WE KNOW THAT ALL THIS WILL HAPPEN?

One of the curious and disheartening things that I have encountered as a practising field scientist is the change in attitude to data. When I was younger there was no question that observations and measurements of Arctic phenomena were automatically accepted as valid, and extrapolations from observed trends were accepted as the best way of forecasting, at least in the short term, what is going to happen. That no longer seems to be the case. If a forecast based on observations gives a result which appears alarming to scientists who look mainly at models, then some scientists seem to ignore it, and may replace it with predictions made by a computer model which may have already failed. I first came across this phenomenon in 2012 when I testified to the House of Commons' Environmental Audit Committee about the rapid decline of Arctic ice, only to be directly contradicted two weeks later by Dame Julia Slingo, Chief Scientist of the UK Meteorological Office, who went out of her way to assure the committee that her modellers said that sea ice would last a long time yet, and ruled out the disappearance of summer Arctic sea ice within the next few years. Again, in 2014 I testified to the House of Lords' Select Committee on the subject of the rapid decline of Arctic ice, only to be directly contradicted by a modeller sitting beside me, who said that models predicted that the summer ice would last until 2050–80. The extraordinary thing is that even a lay person looking at the curves of Plate 16, based firmly on data, can see that there is no possibility whatever that the summer ice can last that long. Yet the advice of such modellers, when given to policymakers, has helped to paralyse them into inaction in the face of a climatic catastrophe which is bearing down on us like an express train.

The trend in the PIOMAS data effectively gives us a drop-dead date of about 2020 for summer sea ice. Anyone who wishes to deny this date and replace it by a much later date must show why the ice volume should deviate above the trend. It would have to do so to achieve a longer-term survival for September ice than a year or two from now, yet there is no mechanism in sight to make this possible. If you do not deny this date, but accept the PIOMAS data as the basis for a best prediction, then this projection does not just lead to an ice-free September in 2016 or 2017, but an ice-free July–November before the 2020s. What is dangerous for the world is that the deniers of this trend comprise not just the normal suspects, such as misguided government scientists or bought-and-paid-for fossil-fuel supporters, but also a body that was set up in 1992 with great hopes of being able to offer a science-based warning of what awaits us if we continue to increase our CO_2 emissions. I mean here the Intergovernmental Panel on Climate Change (IPCC), whose 2013 Fifth Assessment Report (AR5) signally fails to give warning of the early demise of Arctic ice but instead adopts a 'consensus' view that it will be much later this century before the ice disappears. This consensus involves consciously ignoring the observational data in favour of accepting models that have already shown themselves to be false.

This is a serious charge to make against a body for which most scientists have great respect, but it is justified if we look at the Summary for Policymakers of the 2013 Assessment Report,[5] and specifically at Figure SPM.7 on page 21. Figure 7.2 shows part (b) of this figure, and there are four ways in which it is misleading. First, there is a big black bar down the centre at the year 2005. A normal person would assume that the black curve, with its grey error bars lying to the left of that bar, represents historical data on sea ice extent in September since that is what the figure label says. After all, it covers the period from 1950 to 2005 which is safely over and all the data digested. But in fact it is the 'modelled historical evolution using historical reconstructed forcings'.

Northern Hemisphere September sea ice extent

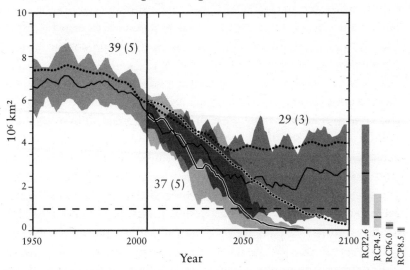

Figure 7.2: The IPCC AR5 Summary for Policymakers, figure SPM.7(b).

In other words, even when the data are available the IPCC prefers to use a historical model, no doubt because it shows a more gentle decline in ice than reality. Stopping the historical curve at 2005 is seriously misleading, because it was from 2007 that the most catastrophically rapid decline of sea ice occurred, and this should not be omitted from the graph. The AR5 assessment is supposed to take account of data published up to 2012, and sea ice extent up to that date is certainly a published quantity. But somehow, in this graph, history stopped in 2005, the transition date for the *previous* AR4 assessment published in 2007. Moving on to future projections, although those projections start in 2005, nine years ago, we see two curves, each with error bars. One is a projection for the 'RCP8.5' scenario for future carbon emissions, and the other is a projection for the 'RCP2.6' scenario.

I need briefly to explain what this unnecessarily complex new way of looking at greenhouse gas forcing is. RCP stands for 'Representative Concentration Pathways'. The number represents the approximate total anthropogenic radiative forcing in the year 2100 relative to that in 1750 before the Industrial Revolution. So 8.5 is 8.5 watts m^{-2}, which is generally agreed to be the level we will reach on a 'business-as-usual' scenario where we don't do anything much about carbon emissions (the world is following, and even exceeding, this scenario at the moment). RCP2.6 is the shameful projection, since it foresees 2.6 watts m^{-2} for 2100, a state which we will pass by about 2030. Why was this included when it is absolutely impossible for us to attain it, however virtuous we become? The anthropogenic forcing reached 2.29 watts m^{-2} in 2011, having risen from 0.57 in 1950 and 1.25 in 1980. Its doubling time seems to be about thirty years, and there is no way that we can bring this under control so that it is only 2.6 by 2100. The RCP2.6 is thus a wholly misleading figure, which looks like it might have been entered into the analysis purely to lull the reader into a false sense of security, into the feeling that if we try hard we can bring warming under control so that the comfortable-looking projections prevail rather than the nasty ones. The IPCC has already admitted that the RCP2.6 scenario can only be achieved by *taking carbon out* of the atmosphere, using methods yet to be invented, rather than just reducing our carbon emissions (which we don't seem capable of doing anyway).

Let's return to figure SPM.7 and its projections. Both projections are highly dubious. RCP8.5 shows the summer sea ice extent declining steadily and reaching effectively zero (that is, dipping below 1 million km^2) by 2050. Yet the curve starts in 2005, where we have already shown that it avoids the embarrassment of having to be compared with real data. In fact in 2012 the real September area had already dropped to 3.4 million km^2, while the RCP8.5 scenario shows the area reaching this value only in 2030. We are already there! Why, then, show a model which includes no real data at all? And let's remember that this purports to show the high-emission scenario. The impossible low-emission scenario, RCP2.6, shows the sea ice extent *never* reaching zero and actually *recovering* later in the century so that it stands at a respectable 3 million km^2 in 2100, not much less than it is today. Where did that clever dodge come from? When this was first published I was telephoned by two reporters who, having seen these curves, said, 'Oh, I see that IPCC predicts a sea ice recovery this century. This means that we don't have to do anything about global warming, doesn't it?' The figure drafters certainly achieved their purpose. It is a brilliant exercise in scientific legerdemain.

The fact is that the 'consensus' who hold the balance of power in the IPCC can't answer the question that I asked at the beginning of this chapter, because their models can't even explain where we are today, let alone where we are going in future. And it's not about being reasonably able to prove that something will happen. Given the existing data, the burden of proof rests with the deniers. We are not even talking about a precautionary principle, as in the case of methane, where we should take action just in case. The case here is solid and should be the basis for action, not something to be denied and hidden. There will be a terrible price to pay if an absurd 'consensus' leads us to ignore the rapid changes which are occurring, and their implications.

IMMEDIATE CONSEQUENCES OF ICE RETREAT – NAVIGATION IN THE ARCTIC

Clearly the future of the Arctic is one of greatly reduced ice cover, especially during the summer months. In the next chapter we will show that this has enormously important implications for the climate system, with potentially disastrous consequences following from the feedbacks set in motion by this ice retreat. However, there are also ramifications for two of Man's everyday commercial activities, shipping and oil exploration.

Shipping in a more ice-free Arctic involves three new possibilities: commercial use of the Northwest Passage across the top of America; commercial use of the Northern Sea Route across the north of Russia; and possible development of a genuine transpolar route from Bering Strait to Fram Strait.

As I mentioned in Chapter 1, trying to traverse the Northwest Passage has always involved a battle against the ice. The quest for the Northwest Passage involved early explorers in two impossible tasks at the same time – investigating the extraordinarily complex web of channels that lay between Baffin Bay and Bering Strait, and doing so during the very short season in summer when the ice weakens and breaks up enough for a sailing ship to be able to make some progress. The two tasks cannot be accomplished together, but one of the reasons why the Royal Navy made protracted efforts was the fact that one of their first attempts was very nearly successful. In 1819 Lieutenant (later Admiral) William Edward Parry was sent out in the *Hecla* and *Griper* and succeeded, by pure luck and the exceptionally easy navigation season, in sailing right through Viscount Melville Sound and getting to Melville Island, where he spent the winter and then returned. This represents almost a complete transit of the Passage. He was never able to repeat this feat in later expeditions, and nor was anybody else able to match it, not through lack of skill but because of too much ice. Navigation in the Northwest Passage is very variable from year to year because for it to be possible the ice has to break up in summer and the broken pieces then have to be swept by wind and current out of the vital connected channels that permit a transit to be made. Parry came closest, but nobody in the nineteenth century ever encountered a truly ice-free Passage, and so the Northwest Passage was never navigated in a wind-powered ship.

With the advent of steam power for ships this should have become easier, but the first steam engines fitted to Arctic exploration ships were very weak and could only be used for short periods because of high coal consumption. In 1845 the Admiralty sent out Sir John Franklin on an expedition which was supposed to solve the Northwest Passage problem once and for all. His ships, *Erebus* and *Terror*, each had a railway locomotive in the hold, connected to a screw by a glorified rubber band, but they were only of 25 horsepower, scarcely enough to move the ships through the water (maximum speed was 4 knots), so it is not surprising that they did not prevent him from getting stuck off King William Island. Tragically, after Franklin had died (probably of natural causes), his second-in-command abandoned the trapped ships and led the crew on a hopeless overland trek to the south; all 128 died. The Northwest Passage was finally successfully achieved by Amundsen in 1903–6. Amundsen, as always, brought Scandinavian ability and common sense to the problem, and used a small, single-masted herring fishing boat, the *Gjøa*, equipped with a hot-bulb engine, an early form of petrol engine. The main advantage of this ship was her small size and draft, which enabled her to sail close to land in the shallow water gap which exists in summer between the grounded broken ice blocks and the shoreline. Because of his ships' draft, Franklin had to stay out in mid-channel and thus got stuck. Amundsen stayed for two winters on King William Island, in a place now known as Gjoa Haven, to take magnetic measurements and commune (in more ways than one) with the local Inuit population in order to learn their techniques of travel, clothing and hunting. It was his Arctic university.

After Amundsen the Northwest Passage was almost forgotten. It had been achieved but it was clearly not a practical shipping route. The next ship to traverse it was the Royal Canadian Mounted Police motor schooner *St Roch* on patrol in 1940–42, with the legendary Sergeant Henry Larsen in command. Then, after the War, the big ships finally started to pass through; first the military icebreaker HMCS *Labrador* in 1954 (I had the enormous pleasure of sailing aboard her in 1978 for a sea ice study off Newfoundland, not long before she was scrapped). Then, moving up in size, came the huge tanker *Manhattan*, 105,000 tons deadweight, which was envisaged by its owners, the misnamed Humble Oil Inc., as the way to transport Arctic oil out of Prudhoe Bay on the north coast of Alaska to markets in the east and Europe. She had a specially strengthened bow but not enough power for her size. She got stuck several times and had to be broken out by the powerful Canadian government icebreaker *John A. Macdonald*. She also developed a leak from ice impacts, and the fresh water in her tanks gradually became salty as a result. Her two voyages were in 1969 and 1970, and as a result of this negative experience the Trans-Alaska Pipeline was built as an alternative, but very expensive, way of getting North Slope oil to market.

In 1970 too it was our turn to attempt the Northwest Passage in CSS *Hudson*, accompanied by our near-sister CSS *Baffin*, during the 'Hudson-70' expedition with which I began this book. Our Captain chose the direct, northerly route through Prince of Wales Strait and into Parry Channel, the course taken by the *St Roch* in 1940, the *Labrador* in 1954 and the *Manhattan* in 1969, but north of Amundsen's passage through Peel Sound, Franklin Strait and Coronation Gulf. We had an easy passage up Prince of Wales Strait, but at the northern end were trapped by impassable ice at the southern end of M'Clure Strait, on the western end of Parry Channel. This has historically offered a frequent blockage of the Passage. True polar ice from the Arctic Ocean can move down through M'Clure Strait and produce a complete barrier of heavy multi-year ice. This was what sank M'Clure's own ship, HMS *Investigator*, when she was searching for Franklin in 1855 – the wreck of the ship was discovered as recently as 2013 by Canadian divers. Like many of our predecessors, we had to be broken out of our ice prison by the *John A. Macdonald*, which came to our assistance so that we were able to arrive safely in Halifax in time for our official welcome as the first ship to circumnavigate the Americas.[6]

Shipping now became more common in the Passage but there was no suggestion of a cargo shipping route. Government icebreakers made the journey, the occasional adventurous yacht slipped through, and there was an effort to establish a cruise ship route through the Passage, in which I played a small role as a tour guide. My ship was the ice-strengthened *Frontier Spirit*, quite a small ship of only 6,000 tons. In 1991, trying to sail west–east, she got no further than the north coast of Alaska. The next year, sailing east–west, she did better and got through, but did require icebreaker assistance from two Canadian government icebreakers, the *Terry Fox* and the *Franklin*. She got stuck just off the west coast of King William Island, in almost the exact spot where Franklin's ships became trapped in 1845 and ultimately wrecked. Again, this was because Arctic Ocean ice from M'Clure Strait had moved even further southeast and was delivering some very heavy, old multi-year ice to the area.

Even when the Passage experienced the milder ice conditions of recent years it did not automatically become navigable. As we see from the maps in Plate 13, it was fully open in 2007 but not in 2005. Even though there is much less ice remaining in summer, and even though the ice always breaks up, it is still not guaranteed that the wind and current will drive all the broken ice blocks out of the Passage and leave a free run. For this reason, I suspect that through-passage cargo shipping will take several more years to get going, although ore carriers already frequent the eastern end of the Passage as they serve iron mines in Baffin Island. There are also plans to send a large cruise ship, *Crystal Serenity*, through the Passage in 2016 after a smaller cruise ship, *The World*, was successful in 2012.

By contrast, the Northern Sea Route (NSR) north of Russia is proving an economic success story. The geography is much simpler; all that has to happen is for the ice to retreat some way northward in summer to permit an uncluttered passage close to the mainland coast. The main blockage today is the Vilkitsky Strait north of Siberia, where the coast takes a northward turn and the New Siberian Islands get in the way, sometimes holding ice around themselves through the summer. When we look at Plate 13 we see that the Northern Sea Route was ice-free in 2005 but not in 2007. In more recent years it has been ice-free every summer and adventurous shipping companies have started sending cargo ships and tankers all the way through. In 2013 there were 49 transits during a 154-day season with 1,355,897 tons of cargo distributed among ports at the eastern and western ends of the sea route.[7] In 2014, because one or two regular shippers dropped out, the tonnage fell to 274,000. Nevertheless there seems to be bright promise for LNG (liquefied natural gas) carriers bringing Arctic gas to market; oil tankers doing the same with oil; cargo ships serving Siberian communities; and various specialized ships. For instance, it has been suggested that Japanese freezer ships that buy salmon and other fish from US fishermen in the Aleutians could deliver it straight to Europe via the Northern Sea Route; and it has been proposed as a route for transporting spent nuclear fuel so that the ships can avoid any chance of being attacked by pirates. But, curiously, it is not seen to be very promising for container shipping, because the pattern of container trade requires a number of intermediate stops between load point and destination, which is not possible in the Northern Sea Route. That has not prevented enthusiastic local authorities from Orkney (Scapa Flow) and Iceland from proposing their towns as sites for great Arctic container terminals. None of this is new: portions of the NSR were in use between the wars for transporting wretched political prisoners to the most hideous parts of the Gulag Archipelago, and there was also a vigorous trade to ports along the NSR which required part-passage only. Many years ago, after I gave a lecture in Cambridge, a British ex-merchant seaman came up to me and said that he had served on British timber ships sailing to Igarka during the 1930s. What is new is the reliability of shipping in the summer because of the high probability that the whole route will be ice-free.

Thus, we already have one reliable trans-Arctic shipping route in summer, with a second on the way. The ultimate goal, which depends on a further retreat of the sea ice, will be a true transpolar shipping route, taking ships from the north Pacific through Bering Strait then directly across the North Pole to the Atlantic via Fram Strait. Savings will be enormous: Yokohama to Hamburg is 6,600 nautical miles through the Northern Sea Route and 11,400 nautical miles via Suez, and the saving on a direct transpolar route will be even greater. The other advantages of a transpolar route are deep water over most of the route, and independence from regulatory authorities and their fees, particularly Russia. There will still have to be safety regulations, and also arrangements for Search and Rescue (SAR) in the event of an accident, which are being organized by the Arctic Council, the association of the eight nations that possess Arctic territory (Russia, the USA, Canada, Sweden, Finland, Norway, Denmark and Iceland). Ice-strengthened vessels capable of carrying cargo through first-year ice without icebreaker escort are now being enthusiastically designed by shipbuilding nations such as South Korea. Such ships could be similar to the *Norilsk Nickel*, an Arctic nickel carrier with an ice-strengthened stern which proceeds stern-first in ice with the aid of an Azipod drive that can rotate through 360°.

IMMEDIATE CONSEQUENCES OF ICE RETREAT − OIL AND THE SEABED

Another immediate consequence of sea ice retreat is that the Arctic is more open for oil exploration than in the past. Until recently most oil exploration has been carried out in shallow water. In the Beaufort Sea, for instance, the earliest offshore wells were in very shallow water, only a few metres deep, off Prudhoe Bay and in the Mackenzie River delta, and were constructed by simply piling up sand into a berm and putting the drill rig on top – an artificial island. Then the search moved into deeper water, a few tens of metres deep, but could still be handled by some form of bottom-mounted structure. In the Russian Arctic too, drilling off the Yamal Peninsula and off Sakhalin was done in water tens of metres deep using bottom-mounted platforms. These shallow waters were part of the seasonal ice zone, ice-free for a part of the year.

But then the search for oil, and the ideas about oil occurrence, spread to deeper and deeper water. In non-Arctic areas this led to drilling in very deep waters off Brazil and, disastrously, in the Gulf of Mexico where the *Deepwater Horizon* tragedy occurred in 1,800 metres of water. The oil industry has now cast its eyes on the deeper waters of the Arctic, beyond the very shallow and well-defined continental shelves. But here industry and politics meet. The Law of the Sea for the Arctic Ocean has still not been agreed. In principle, the area beyond 200 km from the coastline is international waters, coming under the jurisdiction of the UN Seabed Authority. If the continental shelf extends further (as it does with very wide Arctic shelves), then the nearest coastal state can extend its jurisdiction out to the shelf break but no further. Any further claim is subject to intense scrutiny.

The trouble is that the Arctic possesses one feature which can be the subject of endless legal arguments, the Lomonosov Ridge (fig. 7.3). This ridge begins north of the Greenland–Ellesmere Island boundary and extends across the Arctic Ocean, passing near the North Pole, until it reaches the Siberian continental shelf. It extends out from the Siberian shelf, so Russia claims it. It extends out from the Canada–Greenland boundary, so both Canada and Denmark claim it. But most other countries say it should be international. The Ridge itself is actually a fragment of Siberian continental rock that was split off when the Arctic Mid-Ocean Ridge opened about 80 million years ago and started to create new ocean crust, pushing the Lomonosov Ridge away from Siberia. It has now reached the middle of the Arctic Ocean. It is not in fact connected to Siberia, or to Canada or Greenland, so all three claims deserve to fail, since the ends of the ridge are of different rock than the shelves that they grate against. Nor is it really a shelf; it was part of a shelf once, but a quite different piece of shelf than the area that it finds itself in today. Really it should be under UN jurisdiction, but Russia, Canada and Denmark are intent on making national claims, in support of which Russia carried out the infantile performance of dropping metal flags at the North Pole from a submersible in 2007, in 4,200 metres of water.

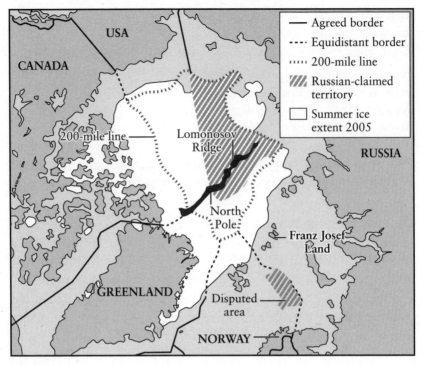

Figure 7.3: The Lomonosov Ridge, showing seabed claims by Russia.

Once the ownership of the seabed is determined, oil exploration can go ahead in deep water, and the retreat of the sea ice will make this much easier. However, the next phase of drilling will be done in tens of metres of water using drill ships, dynamically positioned and protected from summer ice forces by icebreakers endlessly circling around them, breaking up the ice into small fragments that are unable to knock the drill ships off station. Such a summer drilling scenario will be easier if the ice is thinner, or better still non-existent, and the drilling season can be extended in both directions to occupy a larger proportion of the year. If the production stage is reached this can be carried out, as the Russians are doing in the Pechora Sea, using a strengthened production platform feeding ice-strengthened tankers. The idea advocated by climate change scientists that unproduced oil and coal should be left in the ground, since we have already exceeded the acceptable carbon load for the Earth's atmosphere, will be fiercely resisted by oil companies and by politicians greedy for tax revenues. Apart from ideological reasons, the oil industry realizes that a global decision to cease new exploration will instantly devalue every oil company's assets and will thus lead to financial collapse, both of the companies and perhaps of the fragile global financial system.

THE PROBLEM OF OIL SPILLS AND HOW TO DEAL WITH THEM

One widely recognized threat to the Arctic environment is the threat of oil spreading from a seabed blowout. This was the subject of a report by a panel of the National Research Council of the US on which I served.[8] We concluded that if a blowout occurs from the seabed there is no known method of cleaning it up. The oil would rise from the seabed as part of an oil-gas plume and spray the underside of the sea ice with oil droplets which would gather into slicks. The ice underside is constantly moving, so if there is a blowout the oiled ice moves away from the site, while clean ice moves in to be oiled in its turn. In winter, new ice would quickly grow under the oil slick, creating an 'oil sandwich' in which the oil is encased in ice for the rest of the winter. During that time the ice floe may travel 1,000 km or more, ending up in a quite different part of the Arctic from where it started. Then, as surface melt begins in springtime, the oil starts to rise to the top surface of the ice by moving up through brine drainage channels (see Chapter 2), which partially melt and open up in spring, providing a pathway to the top of the ice. Suddenly little patches of oil will sprout up everywhere at the tops of these channels, usually too small to be removed or burned. Later in the summer, when the whole floe melts, the oil will be deposited in the water and so become a very widespread pollutant of the open water summer Arctic. This is especially dangerous for the marine ecosystem and for millions of migratory seabirds.

Most of this knowledge was acquired as far back as 1974–6, during a Canadian Government research programme called the Beaufort Sea Project, on which I worked.[9] The Canadian Government wanted to understand the nature of the oil-under-ice threat before permitting drilling in ice-covered waters, and actually allowed quite large oil spills to be carried out in the Arctic to determine what would happen, including a spill under fast ice which lasted throughout a winter. I well remember one offshore experiment in the Arctic Ocean where we were pumping oil under a pressure ridge, with divers to follow its progress. I was the designated pumper, and worked away with a hand pump, spraying crude oil under the ice but also over my parka to the point where I had to throw it away afterwards as the stench could not be removed. Since then the progress of science has been glacially slow, because political correctness decreed that oil could not be spilled in the Arctic, even for experimental purposes, so concern for the environment prevented any progress being made in determining just what effect an oil spill will have on the environment. When we were working on the 2014 US report we found to our amazement that the 1974–6 Canadian project was still the best source of data.

Our conclusion in 2014 was that a big blowout under ice would actually be more devastating than the *Deepwater Horizon* blowout, because of the way that the ice ensures that the oil is distributed widely around the Arctic Ocean at low concentrations, which make clean-up difficult. We also concluded that drilling a relief well (a commonly recommended method for stopping the blowout) would take too long, so each operator should have a capping device available to put over the blowout and quell it quicker. The first victim of this new view was Shell in 2012, which built a capping device only to have it collapse when first tested. Shell persisted in their plans and started to drill in the Chukchi Sea in 2015, only to abandon the operation in the first season. The NRC panel is hopeful that our conclusions will be accepted by regulators in defining how Arctic drilling should be done. The creation of the panel was motivated by the fear of an Arctic oil rush in which environmental protection would be forgotten in the hurry to stake out new sites and produce oil as a consequence of retreating ice. But this has not happened. Companies have been very cautious. The likely reason is that the massive costs of the *Deepwater Horizon* disaster had to be carried by BP (an estimated $54.6 billion in fines, clean-up costs and settlements) and almost bankrupted them. The polluter pays. And if an Arctic blowout occurs, especially if it is in US territorial waters, the costs will similarly be carried by the polluter and could well be even higher than the Gulf of Mexico. Under these circumstances oil companies have held back and find it more attractive to buy into the fracking boom.

The fear among both industry and regulators of an expensive accident has led to some surprising decisions. The recently rejected Canadian Government of Stephen Harper was not famous for its concern with environmental issues, having been responsible for firing a large number of federal environmental scientists while pushing the expansion of Alberta tar sands, one of the most wasteful possible forms of fossil fuel because of the energy required to 'cook' it so as to extract useful hydrocarbons. Yet on 2 April 2014 we heard that Canada's Federal Transport Minister, Lisa Raitt, had come out bluntly against shipping oil through the north of Canada. Canada has an Arctic port on Hudson Bay at Churchill, Manitoba, which is linked by rail to southern Canada. A plan was mooted by a company called Omnitrax Ltd to send oil by rail car to Churchill then export it by ship through the eastern half of the Northwest Passage to Europe. The minister said:

> I can tell you: one oil spill or accident in the Arctic is one visual you do not want to have in this world at all . . . It's not just always about the economy. I can't believe I said that as a Conservative. But it's not always about the economy. You've got to balance it out with what's happening in terms of safety, and the environment too.

The Manitoba provincial government had been seeking to make the shores of Hudson Bay a protected area, partly because they are at present a refuge for polar bears who, driven from their normal habitats by global warming, congregate in Churchill to feast out of rubbish bins. But they also sought to protect the rare beluga whales which use the nearby coastal waters. It was unusual, but heartening, to see the Federal government supporting this stance.

The retreat of the Arctic sea ice ensures that a year-long blowout will indeed end in tears. Under the blowout scenario envisaged in the 1970s, oil that has been sandwiched and transported round the Arctic is released in summer from floes which melt at the ice edge, giving a fringe of floating oil around the edge of the summer ice. In the future there will be no ice edge in summer because there will be no ice. The oiled ice will melt completely and produce spills which can spread around the whole of the completely open Arctic Ocean. The damage, and cost of clean-up, will be enormous.

I should end by mentioning a closely related question, that of the changing marine ecology of the Arctic as the ice retreats. There is a possibility of new fisheries, given the greater light levels in the water column in spring leading to greater, and earlier, plankton production. It is difficult to predict what exactly the changes in marine ecology will be, but it is certain that the retreat of sea ice will permit fishing vessels to extend their activities geographically and seasonally, so as to exploit whatever living resource is available.

LIKELY COURSE OF FURTHER ICE RETREAT THIS CENTURY

It is extremely difficult to find predictions by modellers as to how the dates of the open water season are going to change as the century progresses. The main reason is the dismal failure of most models to reproduce the present state of the Arctic ice in summer. The debate over exactly when the September Arctic will become ice-free has distracted attention from the more important issue, which is how fast, and in what manner, the Arctic sea ice will retreat at all seasons of the year. The Death Spiral has shown us that within a very few years of the September ice cover disappearing, the ice-free season will have widened to about five months, covering basically July to November. But will it stop there? The Antarctic has just such a seasonal ice cover, with most parts of the Southern Ocean ice-free for four to five months then covered with first-year ice for the rest of the year. Is this stable? Warmer conditions will certainly lead to the ice-free season expanding because of solar radiation combining with warmer air temperatures, but it is likely that even after an ice-free summer, during which ice-free water warms up, there will come a time of year, probably in December, when the combination of darkness, colder air temperatures and the release of stored summer heat from the sea surface will allow ice to form again, and it will stay until the following spring or early summer.

It is difficult to imagine an Arctic Ocean that is ice-free all the year, although an ultimate state like this is foreseen for winter months in the Death Spiral – their ice volumes too are spiralling slowly inwards. An Arctic which is ice-free even in midwinter would develop a completely different water circulation and thermal cycle than a seasonally ice-covered Arctic. This may come about within a century, but by that time far more drastic changes will have happened to our planet which may well have made it uninhabitable for humans. I look at these changes, caused by or related to the retreat of Arctic sea ice, in the next chapter. And we must recognize that much of the damage has already been done; the Siberian shelves are already ice-free in summer and this is producing the threat of a massive methane release, as I shall describe in Chapter 9.

8

The Accelerating Effects of Arctic Feedbacks

THE CONCEPT OF CLIMATIC FEEDBACK

In the last chapter we looked at ways in which the retreat of Arctic sea ice is having direct effects on the future of the Arctic Ocean and on the way in which we view it. At first sight, Arctic sea ice retreat is a boon for the Arctic economically. We can view the Arctic Ocean as a possible trade route, at least in summer, instead of a barrier. We can view it as an easier location to explore for oil and gas, and to exploit marine life. All of these changes are superficially positive, but they are based only on the direct effect of ice retreat in allowing human activities in the Arctic seas during a longer part of the year. We have not considered how sea ice retreat will alter other aspects of the global climate system. This chapter will show that the *indirect* effects of Arctic sea ice retreat are overwhelmingly negative for the planet as a whole, and negative to such an extent that Arctic sea ice retreat has to be seen as an unmitigated disaster for the Earth.

The reason for the difference is *positive feedback*, the fact that Arctic sea ice retreat, directly induced by greenhouse gas warming, has impacts of its own which enhance global change effects on the planet and will cause disastrous consequences out of all proportion to the original change. These feedbacks and connections exist throughout the climate system. As the poet and mystic Francis Thompson put it:

> Thou canst not stir a flower
> Without troubling of a star.

In Chapter 6 we already mentioned one feedback which may be of serious importance, *wave-ice feedback*, whereby ice retreat allows greater summer wave growth in the Beaufort Sea which, interacting with the ice, leads to a greater amount of break-up and melt, and reduced ice growth in autumn. The other important feedbacks that we will consider in this chapter are:

- Ice-albedo feedback
- Snowline retreat feedback
- Water vapour feedback
- Ice sheet melt feedback
- Arctic river feedback
- Black carbon feedback
- Ocean acidification feedback

The potentially most dangerous feedback of all – that of methane release from melting offshore permafrost – will receive a chapter to itself (Chapter 9). Additionally, a further impact has recently become evident, that Arctic ice retreat is associated with, or may even be causing, changes in the position of the jet stream, leading to new patterns of extreme weather at critical times of year in northern hemisphere agricultural areas, creating a threat to global food supplies (discussed in Chapter 10).

ICE-ALBEDO FEEDBACK

We noted in Chapter 2 that the albedo of open water, that is the fraction of incoming solar radiation that is reflected back into space directly, is only about 0.1, while that of sea ice can vary from about 0.5 right up to 0.9. Fresh snow falling on smooth sea ice has an albedo of 0.9, and if you are out on such a surface in March or April when the sun is relatively high and the day long, the dazzling reflections can be enough to induce snow blindness. This was a painful condition suffered by many early explorers such as Captain Scott and his men.

As soon as there are any ridges or other angular surfaces in the ice, or as the snow gradually weathers or is driven by the wind into undulating hummocks called *sastrugi*, the albedo drops to 0.8. When spring comes the albedo drops still more: whenever the air temperature rises above 0°C and a small amount of melt occurs to the superficial snow, it becomes a duller white, and the albedo drops. It drops further when the snow is actively melting, giving rise to soggy slush which may incorporate black carbon that has been deposited over the winter but hidden by successive falls of snow. The final limit is when the surface consists of bare ice riddled with melt pools. The melt pools themselves have dark surfaces which preferentially absorb solar radiation and so melt their way more deeply down into the ice – often melting right through to give the ice a Swiss cheese appearance and very little mechanical strength (Plate 18). The area-averaged albedo at this stage can be 0.5 or even less, but the final drop, to 0.1 when the ice disappears altogether, is the biggest change of all.

There has always been a problem in measuring and modelling albedo in summer. We know the albedo of fresh snow (0.9) very accurately, but this does not much affect the heat balance of the Arctic as there is little solar radiation in winter. In summer, when there is the highest level of radiation, we have to estimate what the average albedo is over a composite surface of melting, slushy snow and ice and melt pools, which can change its nature within hours as surface temperatures rise or fall. The first successful modellers of Arctic thermodynamics, Gary Maykut and Norbert Untersteiner in 1971,[1] had to face up to this problem and chose some fairly arbitrary values for summer albedo, but in recent years there has been a strong emphasis on careful field observations by people like Don Perovich of the US Army Cold Regions Research and Engineering Laboratory, which have shown just how much variability there is.[2]

Albedo change thus has two aspects in respect of climate change. With a warmer climate, surface melt starts earlier in the summer, and the decline of the albedo from a snow-covered value (0.8 or 0.9) to a dirty, melt pool-ridden value (0.5 or so) happens earlier and so allows more radiation absorption during the critical midsummer months. But on top of that, and exceeding it, is the simple change associated with sea ice retreat from a summer surface covered with ice, however dirty, to a free water surface. This takes us down from 0.5 to 0.1, so the details of how albedo declines during summer are less important than obtaining good values for the total ice area so that we know how much open water has replaced what used to be ice. We can see from the bite-sized open ocean of 2007 (Plate 13) just how big that albedo loss might be – and an albedo loss is a radiation gain, a gain in the heating of the planet.

How serious is this albedo loss for warming the planet? A study by Kristina Pistone and colleagues at Scripps Institution of Oceanography[3] estimates that the loss of area of summer sea ice between the 1970s and 2012 has caused a decrease in global average albedo equivalent in its warming power to adding a further one-quarter to the amount of carbon dioxide added to the atmosphere by man during that period. This is called a 'fast feedback' because its effect is immediate. The reduction of short-wave reflected energy leads to a worldwide extra radiative forcing, raising global temperatures. This study did away with the problem of measuring actual ground-based albedos all over the Arctic by using CERES, a satellite that measures radiation values directly. They found that the albedo, averaged over the Arctic and over the year, dropped from 0.52 to 0.48 between 1979 and 2011. This does not seem a lot, but it is the equivalent of 6.4 watts per square metre ($W\ m^{-2}$) of extra radiation being absorbed over the Arctic, or 0.21 $W\ m^{-2}$ when averaged over the whole Earth.

SNOWLINE RETREAT FEEDBACK

Warm air over an ice-free Arctic also causes the snowline to retreat. The sea ice-albedo feedback is enhanced by faster spring snow melt in Arctic coastal lands as sea ice recedes, probably due to warmed air masses moving over the coastal areas from the ice-free sea. If we look at the month of June, when solar radiation is at a maximum, then by 2012 a 6-million km^2 negative area anomaly had developed compared with 1980 (fig. 8.1). That is, compared with the late twentieth century, the midsummer snow extent was 6 million km^2 less. This is of the same magnitude as the sea ice negative anomaly during the same period, and the change in albedo is roughly the same between snow-covered land and snow-free tundra as it is between sea ice and open water. Nobody has yet published the calculations for tundra as Pistone and her colleagues did for sea ice, but the similarity of the magnitudes means that snowline retreat and sea ice retreat are each adding about the same amount to global warming. The overall ice/snow-albedo feedback is

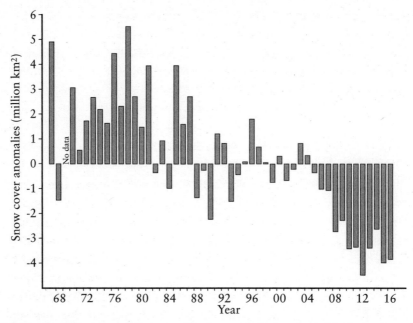

Figure 8.1: Changing area covered by snow in the northern hemisphere during June, 1967–2016.

thus adding 50 per cent (not just 25 per cent) to the direct global heating effect due to CO_2 addition, showing how the Arctic can become a driver of, rather than just a responder to, global change.

This is of utmost importance, yet is not generally realized: when the global feedback arising from the Arctic snow and ice retreats adds 50 per cent to the warming which results from the addition of CO_2, we have reached the point at which we should no longer simply say that adding CO_2 to the atmosphere is warming our planet. Instead we have to say that the CO_2 *which we have added* to the atmosphere has *already* warmed our planet to the point where ice/snow feedback processes are themselves increasing the effect by a further 50 per cent. We are not far from the moment when the feedbacks will themselves be driving the change – that is, we will not need to add more CO_2 to the atmosphere at all, but will get the warming anyway. This is a stage called *runaway warming*, which is possibly what led to the transformation of Venus into a hot, dry, dead world. When Jimi Hendrix played the guitar he had the ability to play passages using feedback alone – his fingers didn't pluck the strings but he manipulated electronic feedback to produce the sounds. We are fast approaching the stage when climate change will be playing the tune for us while we stand by and watch helplessly, with our reductions in CO_2 emissions having no effect.

WATER VAPOUR FEEDBACK

Water vapour feedback is completely dependent on a change in air temperature. We add something like 7 per cent of extra water vapour content to the atmosphere per degree of warming in air temperature, and that in turn adds about 1.5 watts per square metre of radiative forcing, since water vapour is a greenhouse gas. Now the Arctic, with its Arctic amplification factor, is warming fast, while air temperatures globally have been warming slowly in the last decade, probably due to increased heat absorption by the deep ocean. The Arctic thus offers a major local water vapour feedback that severely inhibits the outgoing long-wave radiation and holds heat closer to the surface, that is, to the ice and ocean. If local Arctic temperature has increased by 3°C, as it has in recent years, then the water vapour concentration has gone up by over 20 per cent, adding some 4.5 W m^{-2} to the polar heating over the Arctic basin. This feedback is locally specific to the Arctic, but it is profound and must be included in the total warming effect.

ICE SHEET MELT FEEDBACK AND SEA LEVEL RISE

If albedo feedback is the biggest threat to our continued existence on Earth, the sea level rise associated with ice sheet retreat is also a factor which will make life increasingly uncomfortable for us in the decades to come.

Until the 1980s the conventional wisdom among sea level researchers was that two factors, each of about equal magnitude, contributed to global sea level rise. The first was the warming of the ocean. As the sea warms, because of heat transfer from the warmer atmosphere and greater downward radiation flux from greenhouse gas blockage of outgoing radiation, sea water expands and the surface stands higher. At first only surface waters warmed, but now the warming is spreading to deeper ocean layers. This is called *steric* sea level rise, in which no new water is entering the ocean. The second factor was a rise due to water entering the ocean from land sources, and the main land sources were subpolar glaciers – the Alps, the Himalayas, the glaciers of Alaska and Chile and Norway, and even the few high-altitude glaciers at low latitudes such as on top of Mount Kilimanjaro. We had quite a good idea of how fast these glaciers were losing mass, because glaciologists have made a point of mapping them for many decades, using, initially, traditional methods such as sticking stakes in the glacier and seeing how fast the surface decayed, and, more recently, satellite methods that measure the elevation of the ice surface. It was already the case back in the 1980s that nearly all glaciers in the world were in retreat – we have all seen dramatic pictures of Alpine glaciers which have almost disappeared. I visited the famous Columbia Icefield in the Rocky Mountains in 1970 and again in 2008. In 1970 the glacier snout lay right up against the Trans-Canada Highway, but in 2008 a long bus journey was needed to reach it. Today *all* the glacier systems of the world are in retreat. Figure 8.2 shows the evidence. The few advancing glaciers in the 1980s were themselves products of global warming – the coastal glaciers of Norway were gaining mass because of warmer, wetter winds blowing over them. But now they too are in retreat.

Glacier retreat, plus some other small contributions from dams and hydroelectric schemes, contributes what is called the *eustatic* component of sea level rise, in which water is added to the global ocean. But now it is joined, in fact exceeded, by run-off from the two great polar ice sheets of the world, Greenland and Antarctica. The threat is real. The Greenland ice sheet, with its high latitude and huge elevation of 2–3 km, always used to be solidly frozen year-round, except for a small amount of melt around the edges. Then in the mid-1980s meltwater started to appear on the top of the ice sheet for a brief period each summer. The brief period became longer and the melt area increased. The biggest melt so far was in 2012, when in the period 1–11 July surface melt spread across 97 per cent of the surface of the ice sheet

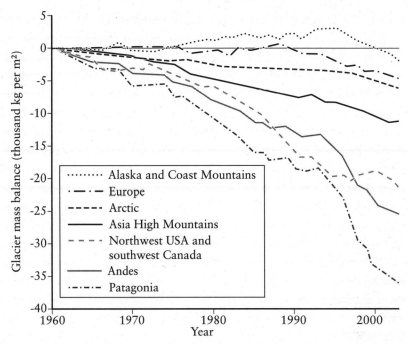

Figure 8.2: Changes in the mass balance of glaciers in different regions of the globe.

(Plate 19). Even then, modellers were not worried. They calculated that most of the meltwater would refreeze at the end of summer so that losses from the ice sheet would be small, and it would take several thousand years for the ice sheet to melt and deposit its water in the sea – which would raise sea level by 7.2 metres. But then a phenomenon developed which they had not expected – *moulins*. These are huge drain holes in the ice sheet surface which penetrate right down through the ice sheet, in many cases to the bedrock 3 km below. The meltwater on the surface can drain through these moulins in a frightening surge. On its way down through the ice sheet the water deposits its heat at different levels, warming the whole ice sheet towards the melting point. At the bottom the water finds its way out to the sea through channels under the ice, lubricating the ice bottom so that the ice sheet, and especially its outlet glaciers, flows faster. Eric Rignot of NASA, using their satellite imagery, has found many Greenland glaciers flowing twice as fast as in the past.[4] This means that they are dumping twice as much fresh water, in the form of ice-bergs, into the ocean. The loss is showing itself in the shrinking of the ice sheet. We can now measure ice sheet mass accurately using a pair of NASA satellites called GRACE (Gravity Recovery and Climate Experiment), which measure very exactly how the mass below them is changing through slight changes in gravity. They have found that the Greenland ice sheet is now losing 300 km^3 of water equivalent per year, a rate which is increasing and which is already as high as the loss from all other glaciers put together.

There are some other minor factors involved in the eustatic sea level rise. One is the transfer of fossil aquifer water into the hydrology cycle. As groundwater is pumped out of underground aquifers, where it has been inaccessible to the atmosphere for many millennia, it is used and then runs off into rivers, evaporates into the atmosphere and is eventually added to the water mass of the ocean. This makes a positive contribution to sea level rise, while other man-made effects, such as water being held back in dams, has a net negative effect because the number of dams in the world is constantly increasing.

Another minor feedback is connected to the change in ice altitude on the ice caps. The overall elevation of the Greenland ice sheet itself is beginning slowly to decrease, and the lower it gets, the higher is the surface temperature of the ice cap (because temperatures are lower at higher altitudes), so the more it melts in the summer. This in itself causes the surface elevation to decline faster, leading to more warm-ing, and so on in a feedback loop. This effect is probably small at the moment, but could become more significant in later stages of decline of the ice sheet, speeding its final demise.

The Antarctic ice sheet has been assumed until recently to be in approximately neutral mass balance, because any melt was offset by snowfall, especially on the mountains around the Antarctic coast-lines. But now GRACE has been applied to the Antarctic, too, and has observed the Antarctic ice sheet also to be definitely in retreat, though not yet as fast as Greenland.[5] The latest estimate is that the Antarctic ice sheet is losing 84 km^3 per year, compared to at least 300 km^3 for Greenland. It is alarming, though, because there is much more ice in the Antarctic to melt, the equivalent of 60 metres of sea level rise. Also, glaciologists calculate that one part of the Antarctic ice sheet, the West Antarctic ice sheet in the Antarctic Peninsula area, is less stable than previously thought, and after a substantial amount of melt could break away from its base. This on its own would produce a sudden sea level rise of several metres.

In the face of these worrying threats the IPCC has been complacent. Indeed in 2007, in its Fourth Assessment Report (AR4), it was seriously so. Because they had difficulty in estimating the eustatic sea level rise, the IPCC authors gave only the steric rise and extrapolated this to the end of the century to give a mere 30 cm of sea level rise by 2100. They pointed out that this was a partial figure that did not include glacier melt, but most non-scientists and policymakers did not read the small print and some very serious underestimates of sea level rise have been used by national bodies responsible for flood defences, for instance the city authorities of Shanghai. The IPCC corrected this fault in 2013 in AR5, but still chose a low figure for their end-of-century estimate (52–98 cm for RCP8.5, the business-as-usual scenario), while most glaciologists believe that the rise will be well above 1 metre, and may be 2 metres. The IPCC figure is based on a linear projection, the assumption that the rate of sea level rise will remain more or less constant throughout the century. But we know that feedback loops lead to non-linear changes. With sea ice disappearance, for instance, the volume of summer sea ice is accelerating downwards in an exponential curve, not a straight line. It makes a big difference. The ice sheet response feedbacks that determine eustatic sea level rise are also exponential processes, or at least accelerating processes, and if we allow for accelerating sea level rise the total amount of rise by 2100 will be much more than linear estimates. James Hansen, former director of the NASA Goddard Space Sciences Institute, estimated that the doubling time in the rate of sea level rise is ten years or less, so the IPCC's complacent predictions of total rise may be seriously overwhelmed in a surprisingly short time.

I joined in this debate in 2004, prompted by a question which has still not been answered. The question was raised, not by me, but by my scientific hero, Walter Munk of Scripps Institution of Oceanography. At that time, just before the GRACE mission made ice sheet calculations easy, an ingenious method of calculating eustatic sea level rise had been proposed by the oceanographer Sid Levitus. He took the census hydrography of the world ocean – that is, the total of all the millions of oceanographic measurements ever made in the world, grouped into a gridded global map – and looked at how the average ocean salinity, taken over all oceans and depths, had changed in fifty years. He assumed that any change must be due to glacier run-off, which would dilute the oceans and reduce the average salinity. His calculations for eustatic sea level rise matched what was known from other methods, so all seemed well. Walter, however, with his trademark scientific insight combined with disarming simplicity, reminded me that I had been measuring the retreat and thinning of sea ice since 1976, and that when sea ice melts it contributes to the dilution of the ocean without contributing to a rise in sea level (this is Archimedes' Principle – the ice is already afloat, like the ice cubes in a gin and tonic). The amount of melting that I was measuring was about 300 km^3 per year, just about the same as the loss due to glacier retreat. Yet the dilution of the ocean measured by the Levitus technique matched observations *without* including any impact from sea ice melt. Why was this? Something was wrong somewhere. Change in ocean salinity equals the contribution from glacier melt, so there is no room for the extra fresh water from sea ice melt. Walter and I wrote a paper together on this anomaly which was published in a distinguished journal.[6] We awaited a response and suggestions from the world sea level research community – after all, Munk is the world leader of oceanography and this was very plainly an anomaly which needed to be resolved. Yet we did not receive a single response or comment on our paper. I even presented it at a global sea level conference in Paris, but the presentation was received without any comments or questions. Walter remarked mildly, 'these sea level people live in a world of their own'. We still await a response to our paper of a decade ago, but the Levitus method has been, in any case, overtaken by GRACE. The only relevant suggestion that I have had, from a physical oceanographer, is that the meltwater from the sea ice may be retained in the Arctic for many years, as it begins to partake in the rotating motion of the Beaufort Gyre, and was hidden from Levitus's averaging technique which was quite poor in the use of ocean data from high Arctic latitudes.

The precise magnitude of the sea level rise produced by a given amount of sea ice retreat has not been determined. Qualitatively, however, we can clearly see that, as the sea ice retreats, warmer air sweeps over the Greenland ice sheet in summer. The summer sea ice has been, in the past, an air-conditioning system both for the atmosphere and the ocean. It kept the ocean water temperature down to 0°C in summer – and, as we will see in the next chapter, the absence of this regulator is having a disastrous effect. It also kept summer air temperatures down near 0°C. Without the moderating effect of summer sea ice, air temperatures over the Arctic Ocean, and thus nearby land, are rising well above 0°C, causing the ice sheet surface to melt.

ARCTIC RIVER FEEDBACK

Another feedback is the warmer run-off temperature of rivers that flow north into the Arctic Ocean. As the snowline retreats on land, the albedo of the land surface in early summer declines dramatically. This leads to a much greater heating of the northern tundra, leading to run-off waters from snow melt that flow through warmer land areas before discharging into the now ice-free continental shelf part of the ocean, so warming it further still and accelerating the ice decline. This in turn accelerates the decline in albedo, which increases the heating of the coastal zone, which drives the snow-line further back, which accelerates tundra temperature change, which increases the run-off heat of the rivers, and so on. The effect is probably less than many of the other feedbacks discussed here, but it is a classic case of a positive feedback developing itself through a sequence of stages.

BLACK CARBON FEEDBACK

A new feedback has been identified recently and found to be somewhat more significant than initially thought – the effect of black carbon deposits from forest and agricultural fires, diesel use and industrial activity on snow and ice reflectivity and melting.[7] This means soot. Glaciologists were used to seeing dirt on glaciers, blown there from surrounding mountains, and found that it is capable of creating an amazing little self-contained ecosystem. A small blob of dirt gathered on a glacier in early summer will absorb solar radiation preferentially, grow warmer than the surrounding ice, and melt itself a little hole into which it sinks. Down in the hole, bacteria can get to work to create a mat of vegetation, with the meltwater dissolving salts from the dirt to provide nutrients. The product is called a *cryoconite*, and I recommend it as an example of how relentlessly life on Earth can establish itself in the most unpromising places. Cryoconites can give a glacier a black, green or even pink tinge.

Cryoconites apart, dirt appears on sea ice during the beginning of the melt season, when the surface snow melts so that the sum of all dirt deposited in the winter appears all together. But until recently this tended to be disregarded, or else included in the calculations of summer albedo. To an extent we can still do this, adjusting our estimates downwards. If we try to isolate black carbon, its global effects seem to be fairly small. The IPCC estimates the radiative forcing due to black carbon at 0.04 W m^{-2}, and observational studies show that its concentration in the Arctic atmosphere appears to have gone down since 1990, perhaps because the worst atmospheric polluters, such as China, are beginning to clean up their operations.

OCEAN ACIDIFICATION FEEDBACK

We know that the ocean is becoming more acidic, and that this is a result of excessive CO_2 dissolving in ocean water to form carbonic acid. The chemical reaction is:

$$CO_2 + H_2O \rightleftharpoons H_2CO_3$$
$$H_2CO_3 \rightleftharpoons H^+ + HCO_3^-$$
$$HCO_3^- \rightleftharpoons H^+ + CO_3^{2-}$$

and there is a complex equilibrium set up between the various ions. The H^+ is the acidic hydrogen ion. As more CO_2 is added to the atmosphere some of it dissolves in the ocean and this acts as a valuable buffer to reduce the rate of global warming. However, the dissolving CO_2 takes part in the above reactions which lead to greater acidity in the ocean, and this will eventually have serious consequences in dissolving the shells of marine creatures (which are made of calcium carbonate), especially the shells of the tiny, single-celled organisms called foraminifera (or forams for short), which are found throughout the ocean (fig. 8.3). The shells of dead forams form a kind of rain in the ocean and are deposited on the seabed where they create a very typical sediment type called an ooze. This is one of the few ways that carbon, added to the Earth's energy system by our fossil fuel burning, can actually be taken out of the system permanently. Therefore if I drive my SUV to the shops, some of the CO_2 that I produce (about 41 per cent) dissolves in the ocean, and some of that is taken up by living forams to grow their calcium carbonate shells. The foram dies, the shell sinks to the seabed, and my CO_2 is harmlessly taken out of the Earth system. The problem is that as the ocean becomes more acidic there comes a point where the shell dissolves again on its long 4,000-metre fall to the ocean bed, because we all know from school chemistry what happens when acid and chalk are mixed. The carbon in the shell is released back into the ocean and stays as part of the Earth system. Even worse, some quite large marine

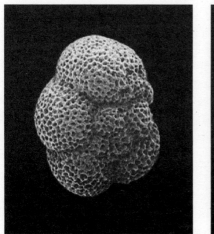 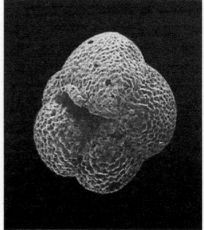

Figure 8.3: Two species of foraminifera from the Arctic Ocean. The shells are tiny, only 0.06–1 mm across.

creatures which grow shells, such as pteropods, will also lose their shells and become shapeless blobs, easy prey for predators. This has been shown in laboratory studies with acidified water. If this is happening we might expect the proportion of CO_2 dissolving in the ocean to diminish, and in fact the latest estimates are that it has fallen from 41 per cent to 40 per cent in thirty years. This is not a large drop, but it is enough to be worried about, especially if we see it starting to accelerate.

Where does sea ice come in here? The retreat of floating sea ice exposes the ocean to acidification because of contact between an atmosphere with more CO_2 and an ocean that has not previously taken it up, so retreating sea ice actually enhances the CO_2 sink. In terms of atmospheric CO_2 this is a negative feedback, bought at the price of increased Arctic acidification. This is a rare case of a negative feedback, although if we consider the extra ocean acidification occurring, and the consequent loss of carbon sink, the feedback in the long run might be positive.

WHICH FEEDBACKS ARE MOST SERIOUS?

If we consider the seven types of feedback listed in this chapter, the most serious is probably the albedo feedback associated with both sea ice and snowline retreat (snowline retreat from coastal lands around the Arctic is itself partly a consequence of sea ice retreat and the warming of winds). If we add the two albedo changes together and include black carbon in the albedo calculation, we get about double the effect described by Pistone and others, that is, albedo feedback is adding 50 per cent to the radiative forcing effect of the CO_2 that we are adding to the atmosphere. It really is equivalent to a case of 'deliver two climate-changing molecules – get one more free'.

The acceleration of Greenland ice sheet melt is also directly associated with sea ice retreat and is leading to a global sea level rise which is accelerating and which will exceed 1 metre this century. Many people think that 1 metre is a minor matter, and that we just need to increase the height of our flood defences by 1 metre. We can do this in the UK, they can do it in the Netherlands and in other rich countries of the world (at a certain cost), but they can't do it in Bangladesh, where 20 million people, mostly poor farmers, live less than 2 metres above sea level. And there is a sinister statistical result that derives from the properties of the bell curve, or Gaussian distribution (fig. 8.4). Suppose the bell curve of fig. 8.4 represents the distribution of sea surface height at a given location, taking into account variability due to tides, winds, etc. The little piece of the curve on the right-hand side represents the height needed to produce a catastrophe – an overflowing of sea defences due to a storm surge, like the one that hit the UK and the Netherlands in January 1953 (flooding my grandparents' house in Tilbury). This area under the curve represents a very low probability. But then let's shift the peak of the distribution by 1 metre, the effect of making average sea level rise by that amount. If we don't raise flood defences, the area under the curve representing disaster

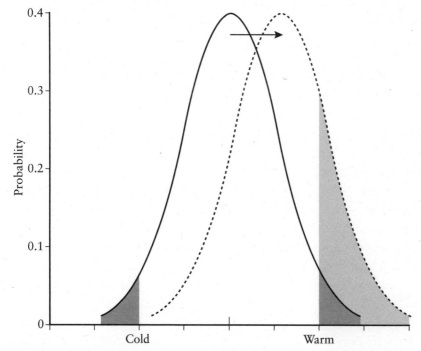

Figure 8.4: Properties of the Gaussian distribution. If the mean is shifted upwards by a small amount, the fraction of occurrences (light grey) which cause disaster is increased by a large amount.

goes up enormously. In other words, a small rise in sea level causes a large rise in the probability of disastrous floods.

Feedbacks show us that Arctic sea ice retreat, when it reaches the levels that we see today, is not just a response to climate change but a driver of climate change. But, of all the threats and dangers that this situation produces, there is one that is potentially even worse – an offshore methane pulse. We consider this in the next chapter.

9

Arctic Methane, a Catastrophe
in the Making

OFFSHORE PERMAFROST AND
WARM WATER

The potentially catastrophic feedback effect that I am about to describe arises from the combination of two phenomena: sea ice retreat, and the continued existence of offshore permafrost in shallow Arctic seas.

I have already described the rapid retreat of the summer sea ice limit, which has removed the sea ice from large areas of the Arctic continental shelves, particularly the seas north of Siberia where there are wide shelves with water depths of only 50–100 metres. What happens to the water column in these newly ice-free seas?

The water structure in the deep Arctic Ocean is made up of three layers. The upper layer, called polar surface water, is about 150 metres deep and is at or near the freezing point. The layer below it, called Atlantic water, extends down to about 900 metres and contains heat which comes from warm North Atlantic water sinking at the ice edge and moving into the Arctic at middle depths. Below that is another cold layer, bottom water, extending to the ocean bed. Therefore if a continental shelf is only 50–100 metres deep there is *only* a single layer of polar surface water present – the warmer, deeper Atlantic water stays outside the shelf break. In the 'old days', before about 2005, this polar surface water layer was covered by sea ice even in summer, and this provided a kind of air-conditioning system in that incoming solar radiation could not warm the water up since its first role was to melt some of the ice. Similarly, the presence of ice kept local air temperatures at about 0°C. From 2005 on, with the summer sea ice completely melted away, solar radiation has been able to penetrate into this shelf water and warm it up. Instead of being held to 0°C all summer thanks to the ice, the polar surface water is now able to rise in temperature through the ice-free summer. In the summer of 2011 a surface temperature of 7°C was detected by a NASA satellite in the Chukchi Sea (as warm as the North Sea in winter). During a recent (August 2014) voyage in which I participated, the US Coastguard icebreaker *Healy* recorded extraordinary sea surface temperatures in the Chukchi Sea; on our way north towards Bering Strait, off Nome, we experienced air temperatures of 19°C and sea surface temperatures of 17°C. Plate 15 shows the extensive warming of surface water in the East Siberian Sea that occurred in September 2007.

Winds over these wide ice-free areas can now create significant waves, which mix the warmed water down to the bottom, so we now have water above freezing point impinging on the Arctic seabed for the very first time in several tens of thousands of years.

On the seabed the warmer water encounters the second element in the story, frozen sediments. These are relics of the last Ice Age, and represent a seaward extension of the permafrost on land. Within them is embedded methane in the form of *methane hydrates* or *clathrates*. This extraordinary solid material looks like ice but it burns. It is a compound of methane gas (CH_4) and water, with an open crystal structure which is stable only under conditions of high pressure and/or low temperature. It is found in various ocean sediments, usually in deep water where the overlying water pressure confers stability. The amount of methane stored in hydrate deposits in the entire ocean bed is estimated to be more than thirteen times the amount of carbon in the atmosphere and amounts to 10,400 gigatons (Gt). On the Arctic shelves, because of the shallow water depth, the hydrates should be unstable, but the solid frozen sediment applies sufficient overpressure to hold them in place. The warm summer water arising from the recent ice loss causes these sediments to thaw out, so they no longer provide a solid cap over the hydrates. The frozen sediments originally formed on land during the Ice Age when sea levels were lower, but then became flooded 7,000–15,000 years ago when the shallow East Siberian Sea formed in the so-called 'Holocene transgression' as the ice sheets melted and sea level rose. So the hydrates, which have rested for tens of thousands of years in the frozen sediment, are now disintegrating as the sediment thaws, producing pure methane gas which is emerging from the sediments and rising to the surface in great bubble plumes. Methane is oxidized in water, so if that rising plume occurs in deep water, as has been seen off the coast of Svalbard in 400 metres,[1] it dissolves and the plume disappears before it reaches the surface. But in water only 50–100 metres deep the methane does not have time to dissolve, and it emerges almost intact from the sea surface into the atmosphere. We must remember – many scientists, alas, forget – that it is only since 2005 that substantial summer open water has existed on Arctic shelves, so we are in an entirely new situation with a new melt phenomenon taking place.

The methane emerges (Plate 21) as huge clouds of bubbles rising through the water column, a process called *ebullition*. The bubbles can be recognized as a series of individual plumes, like plumes from a seabed gas-oil blowout (Chapter 7), originating from a number of point sources on the seabed. The East Siberian Arctic Shelf is exceptionally shallow – more than 75 per cent of its entire area of 2.1 million square kilometres is shallower than 40 metres – so most of the methane gas avoids oxidation in the water column and is released into the atmosphere. Atmospheric concentrations of methane above the sea surface there have been found to be as much as four times greater than normal atmospheric levels. It has been widely assumed that no methane could be emitted from the Arctic shelf during the winter ice-covered period. However, new observational data suggest that methane ebullition and other emissions are now occurring throughout the year. Methane fluxes from European Arctic polynyas have been found to be 20 to 200 times higher in methane than the ocean average, strongly suggesting winter emissions. Methane has also been observed directly accumulating under winter ice. This all suggests that once the cap of frozen sediment has been removed by summer melt, the methane is able to escape at all seasons.

The discovery and observations of these powerful bubble plumes on the East Siberian Shelf in summer were first made by annual US–Russian expeditions led by Natalia Shakhova and Igor Semiletov,[2] which brought back some dramatic underwater pictures (see Plate 20). They estimate that 400 Gt of methane equivalent is held in these sediments, and that 50 Gt could be released from the uppermost tens of metres of the sediment within a very few years of this warming process gathering pace. Modellers such as Igor Dmitrenko[3] at the University of Manitoba have looked at the sediments closest to shore, in only 10 metres of water, and have estimated that the time scales for thawing and methane release are slow, of the order of 1,000 years. But other things are going on further out to sea.

Natalia Shakhova herself, and others,[4] have drawn attention to the role of *taliks* in facilitating methane release from the sediments. Taliks are irregularities in the subsea permafrost layers, caused by faults or localized irregularities, which provide a route for methane to be released from hydrates deep in the sediments and to make its way upwards towards the seabed. Shakhova found that many of the methane plumes observed in the East Siberian Sea consisted of methane being released from the top of a talik. There is a parallel here with the role of moulins on the Greenland ice sheet – they permit thermal processes to occur deeper inside the material than modellers suspected. A talik provides a route for methane molecules to escape from their hydrate cage and move up past the barrier supposedly offered by the seabed permafrost, and then be emitted. The emission rate therefore does not depend on layer after layer of sediment gradually giving up its methane as the permafrost thaws.

Methane is exceptionally powerful as a greenhouse gas. As I said in Chapter 5, it is 23 or 100 (depending how you calculate it) times as powerful per molecule as CO_2 in its warming potential. Arctic offshore emissions may well be the main cause for global atmospheric methane levels starting to rise again in 2008, after stabilizing about the year 2000 (the other likely candidate, leakage from fracking, did not begin until more recently). How much methane is waiting to be emitted in this way, and when will it emerge? What will this do to the climate? We expect that it will further speed up sea ice retreat, reduce the reflection of solar energy and hasten sea level rise as the Greenland ice sheet melt accelerates. But the ramifications of vanishing ice will also be felt far from the poles.

GLOBAL IMPACT OF ARCTIC
METHANE RELEASE

With two colleagues, Gail Whiteman and Chris Hope, I have modelled what a 50 Gt methane release over ten years would do to climate,
both in terms of temperature and cost.[5] Let us remember that, although
this is a huge and seemingly impossible quantity of gas to let loose on
the world (our total annual release of CO_2 is only 35 Gt), it is still less
than 10 per cent of the total volume of methane believed to be locked
into the East Siberian Sea sediments. To quantify the effects of a
major Arctic methane pulse on the global economy, we used the
PAGE09 integrated assessment model, which allows the extra emissions to be traced through to changes in sea level, regional temperatures,
and regional and global impacts, such as flooding, health and extreme
weather, taking account of uncertainty.[6] PAGE09 calculates the
amount by which the Net Present Value (NPV) of impacts, aggre-
gated between now and 2200, increases if one more tonne of CO_2 is
emitted, or decreases if one less tonne is emitted – effectively, the
social cost of CO_2. PAGE09 is the most recent version of the PAGE
model, developed by Chris Hope at the Judge Institute in Cambridge
and used by the Government's Stern Review of the Economics of Climate Change to calculate the impacts of climate change.[7] All results
are based on 10,000 runs of the model, which allows a full picture of
the risks to be built up in order to work out uncertainties.

We tested two standard emissions scenarios. First was the 'business
as usual' scenario, where it is assumed that the world carries on following its present course with increasing emissions of carbon dioxide
and other greenhouse gases year on year without any mitigating
actions. Secondly we ran a 'low emissions' case, with a 50 per cent
chance of keeping the rise in global mean temperatures below 2°C (the
'2016r5low' scenario from the UK Meteorological Office). In each
case, we superposed a decade-long pulse of 50 Gt of methane released
into the atmosphere between 2015 and 2025. We also explored the
impacts of later, longer lasting or smaller pulses of methane.

The extra temperature rise due to the methane by 2040 is 0.6°C, a
substantial extra contribution (fig. 9.1). This would be a catastrophe

for mankind, partly because it is so quick. It would speed up all the other global warming effects and there would be nothing that we could do to shut off the methane except for cooling the water column (i.e. bringing back the sea ice), which is exceedingly difficult to envisage. Such a methane pulse would bring forward by fifteen to thirty-five years the date at which the global mean temperature rise exceeds 2°C above pre-industrial levels – to 2035 for the 'business as usual' scenario or 2040 for the low emissions scenario. Note how rapidly the methane generates its climatic effect, because, although the peak of 0.6°C is reached twenty-five years after emissions begin, a rise of 0.3–0.4°C occurs within a very few years.

Measured at present values the cost of this increase comes out as 60 trillion dollars over a century for the business-as-usual scenario. We had anticipated that the price-tag for changes to the Arctic would be steep, notwithstanding short-term economic gains for Arctic nations and some industries, but it came as a surprise to discover just how steep it might be. The amount is 15 per cent of the total of $400 trillion estimated by the same model as the total cost to the world of

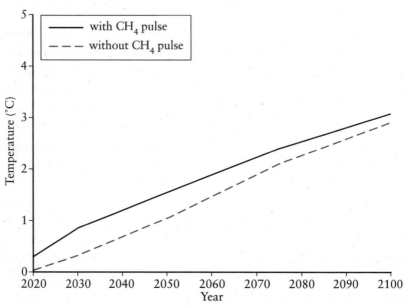

Figure 9.1: Likely change in global average temperature produced by a 50 Gt methane pulse emitted between 2015 and 2025.

all climate change impacts over the same period. For the low emissions scenario the cost will still be an extra $37 trillion. These costs are the same irrespective of whether the methane pulse is delayed by up to twenty years, kicking in at 2035 rather than 2015, or stretched out over two or three decades rather than one. A pulse of 25 Gt of methane has almost exactly half the extra impact of a 50-Gt pulse.

The model divides the planet into eight regions to model where change would result. Under both scenarios, the global distribution of the extra impacts closely mirrors the total impacts of climate change: 80 per cent of the extra impacts by value occur in the poorer economies of Africa, Asia and South America. Inundation of low-lying areas, extreme heat stress, droughts and storms are all magnified by the extra methane emissions. So, a purely Arctic effect, switched on by the global warming-induced phenomenon of sea ice retreat from Arctic shelves, turns out to have a global impact. And, as usual, it is the global poor who will suffer most.

ACTION THIS DAY

When it comes to environmental and climatic change, two criteria are meant to be applied as the basis for action or inaction in the face of a possible threat. The *precautionary principle* states that we should take action to mitigate a plausible threat even if we are not sure that it has happened yet. For instance, when the first assessment of the IPCC was published in 1992, conclusive proof that we were altering the climate by our emissions was not yet available, but the presumption that this was what was happening was sufficiently strong to produce a call for action. *Risk analysis* helps us to quantify the magnitude of the threat. Mathematically, risk can be defined simply as probability of an effect happening multiplied by the negative effect if it does. Particularly difficult to assess are risks where the probability is very small but the effects would be very large, such as the risk of Earth being hit by an asteroid. In the case of an Arctic offshore methane pulse there is no question that the risk is huge. First, the probability of this pulse happening is high, at least 50 per cent according to the analysis of sediment composition and stability by those best placed to know what is going on, Natalia Shakhova and Igor Semiletov. More-over, if it happens the detrimental effects are gigantic, with an economic cost alone of $60 trillion, including high levels of human mortality. So, on any definition, the risk of an Arctic seabed methane pulse is one of the *greatest immediate risks* facing the human race.

Why then are we doing nothing about it? Why is this risk generally ignored by climate scientists, and scarcely mentioned in the latest IPCC assessment? I fear it is a collective failure of nerve by those whose responsibility is to speak out and advocate action. It seems to be not just climate change deniers who wish to conceal the Arctic methane threat, but also many Arctic scientists, including so-called 'methane experts'. For some such experts, accustomed only to minor Arctic methane seeps, just one of many natural and anthropogenic sources of methane, there is some excuse. But they have not woken up to the fact that the environmental conditions now are unprecedented: it is only since 2005 that the Russian Arctic shelf waters have been routinely exposed to the atmosphere over most of their area, permit-ting the water temperature to rise far above melting point. It is perhaps difficult for non-Arctic scientists to grasp that this is an entirely new situation, and that previous concepts no longer fit. Some other scientists do clearly grasp what is going on, yet by a psychologi-cal process of denial prefer to try to wish it away. For example, at a Royal Society meeting on 22 September 2014, methane expert Gavin Schmidt (Director of the NASA Goddard Institute for Space Studies) publicly derided the concept that large amounts of methane could be emitted from the seabed, just when new results from the Laptev Sea were being announced that were demonstrating large increases in emissions. The integrity and accuracy of field researchers have even been called into question, with personal abuse being hurled at Shak-hova and Semiletov because they are Russian and one of them is a woman. This is a pretty low point for the scientific community to have reached, but it has happened in part because the implications of this discovery are truly momentous. Even if we wish to sit on our hands and dawdle when it comes to reducing CO_2 emissions, we can-not sit quietly by while 50 Gt of methane is probably going to be emitted into the atmosphere and cause a rapid rise of 0.6°C in global temperatures. And this is just the first instalment: much, much more methane remains in these sediments and will emerge over coming decades as the sediments continue to thaw, while terrestrial permafrost (see the next section) will add in the long term an even greater amount of methane.

What can we do? For a start, immediate research is necessary on an emergency scale, because there is still too much that we simply don't know. It is true, and easy, to say that if we could halt and reverse global warming in some other way, for example by geoengineering, then the summer Arctic sea ice cover would come back and the shelf water would return to its previous 0°C temperature level. Permafrost thawing and methane emission would cease. But, as methane plumes are coming out now, and already causing radiative forcing, it is difficult to see how we can make a heroic effort to bring down temperatures enough to prevent further methane emission. If we could, we would have already conquered climate change and would have nothing much to worry about. No, the only effective way is directly to prevent the methane from emerging from the seabed sediments under present and near-future conditions, and this is where we are short of solutions. The idea of catching methane in plastic domes or sheets and feeding it to a central flaring site has been suggested, but since the entire seabed seems to be erupting with methane the plastic sheets would have to cover the whole East Siberian seabed, which is impossible. The only suggestion that seems plausible so far comes from the oil industry itself, which is to carry out a version of fracking in which a well is drilled to below the active layer of sediment, with horizontal drilling then connecting the well to cavities created under the sediments. Methane drawn into these cavities could be pumped out and flared. Flaring makes sense, as when a molecule of methane burns it produces a molecule of CO_2 which has only one twenty-third of the heating power of methane. But if the methane can be captured and used, so much the better. This solution would require a network of wells covering the entire East Siberian Sea. Nobody has calculated exactly how many, nor what the entire venture would cost. But in the absence of any other solution, this one must be researched urgently, and if it is practicable we should be ready to implement it. It would be ironic if the oil industry saved the world through its advanced technology, but I am sure that God would raise a smile.

THE THREAT FROM PERMAFROST DECAY ON LAND

The Arctic offshore is our greatest immediate threat, but the threat of methane and CO_2 emissions from decaying permafrost on land is also not only real but inexorable. We know from the careful work of Arctic biologists that as terrestrial permafrost thaws, the rotted surface vegetation which becomes unfrozen goes through a sequence of chemical and biological processes in which it ends up producing both methane and CO_2. This is different from the Arctic offshore permafrost where the methane is already there waiting to be released when the sediments thaw. Here the methane has to be generated by a longer, slower set of chemical processes, but eventually it is still produced.

Let us look at some statistics. The area of terrestrial permafrost in the world today is about 19 million km², including both continuous and discontinuous (that is, patchy) permafrost. It is thawing; permafrost areas have warmed by 2–3°C since the 1980s. When it thaws, permafrost emits a mixture of methane, carbon dioxide and (some) nitrous oxide (N_2O), all of them greenhouse gases. According to the IPCC, the quantity of carbon contained in this permafrost is 1,400–1,700 Gt. It is estimated that 110–230 Gt will be lost (as both CO_2 and CH_4) by 2040, and 800–1,400 Gt by 2100, at a rate of loss of 4–8 Gt per year before 2040, rising to 10–16 Gt per year afterwards.

Please note these figures. What this means is that, by the end of the century, the quantity of carbon that will have been emitted from thawing permafrost on land is some thirty times the 50 Gt offshore methane pulse which we fear in the next decade. It is unclear how much of this carbon is in the form of hyperactive methane, but it is probably substantial. So a major climate warming boost from methane is inevitable – it may be fast, due to the thawing of the offshore permafrost releasing trapped methane; it may be slow, due to methane creation by terrestrial permafrost; or it may be both fast and slow, a pulse from the offshore permafrost followed by a slower but larger release from onshore permafrost. But we are certain to receive this extra boost to warming by the end of the century at the latest.

Once again, an extraordinary aspect of the 2013 IPCC assessment is that these figures on methane emissions from terrestrial permafrost are quoted, but the implications for accelerated climate warming are not pursued, although the implications are as bad as, or worse than, the implications from offshore release.

THE AREA WIDENS

Following the discoveries of Shakhova and Semiletov, the pace of exploration of the Arctic shelves has intensified, yielding further discoveries of warm offshore water and methane production in shelf areas other than the East Siberian Sea.

Semiletov and Shakhova expanded their area of operations out of the East Siberian Sea by joining the Swedish icebreaker *Oden* for the 'SWERUS-C3' cruise to the Laptev Sea in the summer of 2014. The ship found a zone several kilometres across on the outer shelf at a water depth of 200–500 metres where large quantities of methane bubbles were being emitted, while closer to shore they found 100 methane sources on the seabed at depths of 60–70 metres, including one intense methane outbreak at 62 metres which the Chief Scientist, Örjan Gustafsson, termed a 'mega methane flare'. This large emission was discovered on 22 July 2014; it was announced that the team observed 'elevated methane levels, about 10 times higher than background seawater' in the surrounding water column. A borehole through the shelf sediment produced methane.[8]

In January 2016 a report by the Laptev Sea Programme, a Russian–German field research project which has been operating since the 1990s, revealed an extraordinary development.[9] Since 2007 a research station mooring has been in place on the shelf, at a water depth of 40–50 metres, measuring water temperatures from the surface to the bottom and the thickness of the ice. In the remarkable summer of 2012 the instruments recorded an early retreat of the ice cover followed by a warming of the water at mid-depth, driven by heat from the Lena River outflow and penetrating solar radiation. The heat mixed downwards towards the seabed, but took time to do so, so it was *winter* before the seabed water warmed up, to 0.6°C in January 2013, spending 2.5 months at that temperature. This would have a melting effect on the sediments, and links warmer water to the methane observed by 'SWERUS C-3'. Model studies agree that the Laptev Sea could be a larger source of methane than the East Siberian Sea.[10]

The strength of activity in this second Arctic shelf sea leads to the conclusion that methane emission from the seabed is not confined to the East Siberian Sea but is found in more, possibly all, Arctic shelf seas. So our estimates of methane emission are probably still too modest. *In situ* monitoring of the methane levels in the Arctic atmosphere has revealed occasional peaks lying well above background levels (called 'dragon breath' by Jason Box of the Geological Survey of Denmark and Greenland, since each one seems to represent some exceptional emission from a single source). They may originate in individual, unobserved mega flares. The record from a methane monitoring station at Alert, on the northern tip of Ellesmere Island (fig. 9.2), shows that methane levels, which had stabilized at about 1,852 parts per billion (ppb) in 2000, have been rising at an accelerating rate, and have

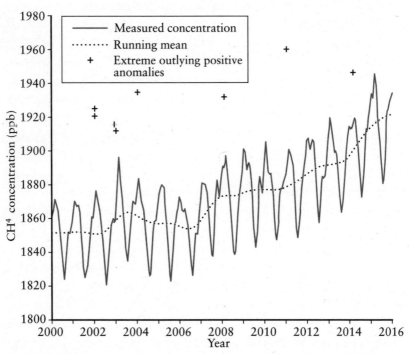

Figure 9.2: Methane levels in the atmosphere measured at Alert, Ellesmere Island, Canada, 2000–2016.

now reached 1,940 ppb, most of the increase having occurred in the last three to four years.

Possibly relevant also is the appearance of three mysterious craters with smooth vertical walls in August 2014 in the north Siberian tundra, surrounded by deposited soil material. The most plausible explanation is that they arose from underground methane explosions, where the melting permafrost allowed a build-up of methane underneath a cap of sediment, and finally blew the cap out in a great explosion.

All of these events strongly suggest that increased methane emission is already under way in the Arctic coastal regions, making use of mechanisms that have never been observed before. It is important that we recognize the threat that this poses to the climate, a threat which is immediate even though it has been belittled by the IPCC in its Fifth Assessment.

10

Strange Weather

The winters of 2009–10, 2010–11, 2014 (January) and 2014–15 brought exceptionally cold weather to the eastern United States and western Europe. This had a serious effect on US maize harvests, with consequent loss of food reserves for tackling famines in Africa. It became clear that, were this to continue for any significant period, disruption of agriculture in the highly productive northern mid-latitudes might create starvation on a mass scale as well as political unrest in vulnerable countries. A single exceptional winter can be regarded as a purely random fluctuation in the ever-changing weather, but when it recurs over a six-year period and begins to look like the birth of a new climate pattern, the question arises as to whether this is related to other observable changes in the Earth system. Because the anomalies are spread out around the world at northern mid-latitudes, and because successive anomalies, moving round the globe, tend to give opposite signals (warm then cold then warm again), a possible cause is a change in the jet stream. A plausible mechanism, first described in a paper by Jennifer Francis of Rutgers University and Stephen Vavrus of the University of Wisconsin,[1] connects a jet-stream shift and weakened zonal (north–south) winds with the loss of sea ice in the Arctic in summer. If this is true, the weather shift would really be a climate shift, with severe economic impacts on mid-latitude economies due to bad winter and spring weather and the increased occurrence of extreme events such as Hurricane Sandy of 2012.

In addition to the impact on the weather, ice retreat may be causing another effect which impinges on our ability to feed the world's population: the loss of ice and snow from mountain glaciers which reduces the spring water supply to crop-producing areas.

WEATHER AND THE JET STREAM

Extreme weather events recorded in recent years are concentrated at northern mid-latitudes. In the winter of 2014–15 alone, San Francisco had its first rain-free January amid continuing drought in California; the Midwest and northeast of the US had extreme cold; and New England had massive amounts of snow.

Northern hemisphere weather is driven by a continuous complex interaction between two great air masses, the polar air mass centred on the North Pole and the warm tropical air mass at lower latitudes. The change in global temperature with latitude is not gradual and even, but instead is rather sudden across the boundary between the lower-latitude and the polar air. This atmospheric boundary, known as the Polar Front, is the collision zone where Atlantic depressions are generated; their track is largely directed by its position. The steep pressure gradients that occur high in the atmosphere, as a concomitant of the boundary, result in a narrow band of very strong high-altitude winds, sometimes exceeding 200 miles per hour, occurring just below the tropopause (the tropopause is the height at which the atmosphere ceases to cool with increasing height and starts to warm again, about 9 km above the Earth in polar regions). Such bands occur in both hemispheres and are known as jet streams. The 'Polar jet stream' usually refers to the one in the northern hemisphere, associated with the Polar Front in the atmosphere. The greater the temperature contrast across the front, the stronger the Polar jet stream: for this reason it is typically strongest in the winter months, when the contrast between the frigid, sunless Arctic and the mid-latitudes is normally at its greatest. Transatlantic air passengers are familiar with the jet stream because it is a powerful tail wind for flights from the US to Europe and a head wind for flights in the opposite direction.

The jet stream is not a straight line, but curves round in great meandering lobes because of the instability associated with the big velocity contrast between the two sides of the boundary. The slower the jet stream, the bigger and slower the meanders. In recent years we have seen an increase in the size of the meanders in the jet stream, that is, the north–south range of the meanderings. This drives another energy feedback: the north-bound air masses on the tropical side of the jet stream boundary bring warmer air into the Arctic, while the south-bound air masses on the polar side take colder air out of the Arctic into lower latitudes than in the past. This increased meandering of the jet stream is, therefore, in itself a heat-transfer accelerator from mid-latitudes to higher latitudes. This in turn then accelerates Arctic warming, decreases the temperature divide between the Arctic air mass and mid-latitude air mass, and thus slows the jet stream and increases the size of its meanders still more, reinforcing the heat exchange feedback. The mechanism identified by Francis and Vavrus can thus properly be called a *jet stream feedback* because the impact of Arctic sea ice retreat upon the position of the jet stream itself feeds back into and amplifies the retreat.

Apart from being larger in amplitude, the meanders move downstream (eastward) much more slowly, causing persistence of weather patterns and thus intensification of events such as drought, flooding, heat waves and cold spells where duration is an important factor.

These enlarged lobes can exert an influence further south than the lobe itself, because of displaced air. One effect that has been suggested is an increase in the frequency of hurricanes in the tropical Atlantic because of the decrease in heat being transported north towards the Arctic (as the Arctic warms up relative to low latitudes, the temperature difference goes down). This leaves more heat in the tropics, and it is warm surface water in the tropical Atlantic and Gulf of Mexico that is the source of hurricanes.

IS THE EFFECT REAL?

The mechanism proposed by Francis and Vavrus is plausible, but it is not the only mechanism by which Arctic sea ice loss can affect mid-latitude weather. In fact James Overland of the Pacific Marine Environmental Laboratory in Seattle has given three reasons why we should be cautious before accepting any direct link between Arctic warming and lower-latitude weather patterns.[2]

First, there is the natural caution which causes scientists not to assign a new phenomenon to a definite cause-and-effect chain until it has recurred sufficiently often to make such an association statistically valid. The enhancement of the Arctic amplification of warming and the significant retreat of sea ice have all occurred within the last ten years, while anomalous weather events have been prominent for six years; a run of six years is not considered enough to distinguish robustly the effect of Arctic forcing from other random events in our weather system. Atmospheric scientists at their mothers' knees are taught the difference between climate – the long-term pattern of atmospheric behaviour – and weather, the local transient effect of a multitude of random causes. The best weather forecasts only show 'skill', that is, only perform better than randomly, out to fourteen days ahead; this is indeed an ultimate limit for weather prediction, since by day fourteen the atmospheric motion has been randomized relative to its state on day one. Scientists who look at images, whether it be the surface of Mars or an atmospheric pressure map, are also warned against *pareidolia*, our innate tendency to look at a random image and see a pattern where none exists. The 'face' on Mars, or the tendency to see a meaning in a Rorschach ink blot test in psychology, are cases of pareidolia. We see the pattern of ice retreat, we see the pattern of extreme weather events, and we believe that one causes the other.

(I had an unrelated personal case of Arctic pareidolia in 1994, when I helped run a private expedition to King William Island to investigate a 'sighting' of Franklin's grave made by an English naval officer. The officer had spent much time searching the west coast of the island, near where Franklin's ships had been beset, and, just as he had run out of food and was about to be airlifted out, he came across a 'burial mound' (actually a glacial drumlin) on which he found a perfect grave shape, a double row of flat square stones forming a pavement seven feet long by three feet wide. Since Franklin's death in 1847 his grave has never been found. Could this be it? With a lot of difficulty and expense we made our way the next year to the site. There it was, a definite grave, looking as solid and grave-like as any tomb in a municipal cemetery. We hammered on the stones and they echoed, implying a hollow beneath. Excitedly, we called in a Canadian archaeologist, who arrived in a Twin Otter aircraft, and who was allowed to lift the stones. In his presence we gently raised the stones and our eager gaze saw … lemming burrows! I have never been more disappointed. After much investigation, we concluded that a more or less perfectly cubic rock, probably deposited by a glacier, had been standing on top of the drumlin, and frost had caused successive thin sheets of stone to crack off the parent rock and slide down the slope, deploying themselves by gravity and accident in a double row that looked like a perfect man-made grave. Lemmings had welcomed the chance to dig homes underneath.)

This, then, is the null argument. There is nothing happening beyond a chance line-up of extreme weather events during six successive winters that happen to correspond with the retreat of Arctic sea ice. If the events continue to follow the same pattern, of course, this null hypothesis becomes less and less tenable.

The second, rather similar, argument is that the jet stream, identified by Francis and Vavrus as the direct cause of the weather events, is actually a very chaotic flow of air which could easily produce random effects resembling something deterministic.

Thirdly, there are other possible forcings of the mid-latitude circulation. The US National Academy of Sciences held a workshop on 12–13 September 2013 which gathered together many scientists with an interest in this problem and led to a comprehensive report.[3] Each scientist had his own pet mechanism. The full list of distinct possible processes is:

- Increased Arctic warming ➔ weakened temperature gradient ➔ weakened, more meandering jet stream and ➔ more persistent weather patterns in the mid-latitude (the original Francis and Vavrus mechanism of 2012)
- Arctic sea ice loss ➔ increases in autumn high latitude snow cover ➔ more expansive and strengthened Siberian high pressure in autumn and winter ➔ increased upward propagation of planetary waves ➔ more sudden stratospheric warmings ➔ weakened polar vortex and weakened, more meandering jet stream (Cohen et al., 2012[4]; Ghatak et al., 2012[5])
- Arctic sea ice loss ➔ changes in regional heat and other energy fluxes ➔ unstable polar vortex ➔ cold polar air moving to the mid-latitudes (Overland and Wang, 2010[6])

- Arctic sea ice loss ➔ a more meandering jet stream and winter atmospheric circulation patterns similar to a negative phase of the winter Arctic Oscillation ➔ frequent episodes of atmospheric 'blocking' patterns (Liu et al., 2012[7])
- Arctic sea ice loss ➔ southward shift of the jet stream position over Europe in summer ➔ increased frequency of cloudy, cool, and wet summers over northwest Europe (Screen and Simmonds, 2013[8])
- Arctic sea ice loss ➔ winter atmospheric circulation response resembling the negative phase of the Arctic Oscillation ➔ rainfall extremes in the Mediterranean in winter (Grassi et al., 2013[9])
- Arctic sea ice loss ➔ negative phase of the tripole wind pattern ➔ enhanced winter precipitation and declining winter temperature in East Asia (Wu et al., 2013[10])

Clearly matters are not as simple as they are for some of the obvious feedbacks that we have been considering. Much more research needs to be done on these mechanisms, but it is conspicuous that *all* the mechanisms described ascribe the origin of the mid-latitude weather events to Arctic change, specifically sea ice loss, and all of them involve feedbacks, though it is only in the Francis and Vavrus mechanism that the feedback is simple and direct.

Two phenomena have been found where there is genuinely strong evidence that Arctic warming and ice retreat can be causally linked to a weather event. The first is in eastern Asia, where the loss of sea ice in the Barents and Kara seas can be linked to the strengthening of the Siberian High (a persistent high pressure area over Siberia), which causes cold air outbreaks into eastern Asia. The second is the case of cold air penetration into the southeast United States, which is related to a shift in the long-wave atmospheric wind pattern, reinforced by warmer temperatures west of Greenland.

WEATHER EVENTS AND FOOD

Extreme weather events have already been found to affect agricultural production, as they impinge on the highly productive mid-latitude northern hemisphere agricultural regions. If there is a causal link between them and the retreat of the Arctic sea ice, with all that follows from it, we can expect these extreme events to become the new annual norm, a revised climate cycle for the Earth. The Arctic sea ice is not going to return of its own accord any time soon, and the continuing increase in greenhouse gas concentration will ensure, via the Arctic amplification, that there will be a rapidly warming Arctic over the next many decades. The impact of extreme, often violent, weather on crops in a world where the population continues to increase rapidly can only be disastrous. Sooner or later there will be an unbridgeable gulf between global food needs and our capacity to produce food in an unstable climate. Inevitably, starvation will reduce the world's population. Scientists desperately hope that there is no link between global warming and this possible change in weather patterns, which is perhaps why they cling to the null hypothesis in the face of increasing evidence to the contrary.

The idea that Arctic change may affect lower-latitude weather is not new, but it has never been expressed before with a mechanism attached. The Arctic as a weather marker for the world was the motivation behind the life and work of Sir Hubert Wilkins, the polar explorer. Wilkins was born on a sheep farm in Australia and saw for himself the disastrous outcomes of drought on farmers' livelihoods. In his book, *Flying the Arctic*,[11] published in 1928, in which he described his pioneer flight across the Arctic Ocean from Alaska to Spitsbergen, he wrote:

> People often ask me why I go to the polar regions . . . From evidence collected many years ago scientific meteorologists deduced the theory that data collected in polar regions and correlated with meteorological information from other latitudes would enable us to forecast the seasons with comparative accuracy. The maintenance of polar meteorological stations during recent years has proved that there is a direct relationship between the Arctic, the Antarctic and subsequent conditions in the great producing areas of the world.

It is these 'great producing areas' which appear to be most threatened by the polar-linked changes to jet-stream location. And some countries can apparently see what we fail to see in the West and take self-protective action. China, for instance, has been buying up or leasing agricultural land around the globe, primarily in South

America and Africa. The industrial agricultural practices which they introduce lift a small number of farmers out of poverty while impoverishing the rest. In the long run they also damage the soil, biodiversity, drinking water and river and ocean habitats. But China is positioning itself for the struggle to come, the struggle to find enough to eat. By controlling land in other countries they will control those countries' food supply.

The effect of climate change on food production can be seen in the Food Price Index (FPI), an international measure of global average food prices maintained by the UN Food and Agriculture Organization. Taking 2002–4 as 100, food prices rose rapidly from 2004 onwards until they reached 230 in 2011, since when they have descended to 150 (2016). If we compare the index with political events we can see that the huge rise was closely followed by the Arab Spring of 2011, which began as a protest against food prices and the way they hit the urban unemployed. It is almost always the case that peaks in the FPI are associated with social unrest in those Third World countries where the cost of food is a large fraction of a person's outgoings. Food cost rather than the absolute absence of food can often be the key factor in shortages and possible starvation. During the height of the Irish Potato Famine in 1845, Ireland was actually exporting food to England. The peasants starved because they could not afford to buy food at the local prices, enhanced by the loss of the potato crop. There was enough food, in absolute terms, to keep everyone alive; they died because they had no money to buy it.

As well as 'natural' factors (themselves man-made when traced to their origin) such as extreme weather events, we are deliberately making our food situation worse by turning human food crops into biofuel. The most notorious case is maize, which is not only a staple food but is also the basis of US food aid to Africa in times of famine. President George W. Bush enthusiastically embraced the idea of turning maize into biofuel, which is now what happens to 40 per cent of the US maize crop. As might be expected, this has led to a collapse in the global food reserves which are available to relieve famine. The European Union was all set to follow the US, until a timely report in 2012 by a committee of the European Environment Agency,[12] of which I was a member, demonstrated that biofuel is not even efficient in terms of greenhouse gas reduction, let alone being good for human food supply.

In summary, we know the following:

1 Weather patterns in the northern hemisphere in winter and spring have changed in a noticeable way, with greater prevalence of extreme conditions.
2 This has led to disruptions in food production at a time when the human population is expanding fast, and is associated with a rise in the Food Price Index which, if it resumes, could lead to fresh rounds of food deprivation and civil unrest in countries which have difficulty feeding their populations.
3 If the mechanism is indeed connected with the loss of sea ice in summer then we cannot expect a natural amelioration.

THE PROBLEM OF WATER

Intertwined with the problem of securing sufficient food for the human race is the problem of water supply. It was drought in Australia that inspired Hubert Wilkins' life's work on polar climate. Globally, as the population increases, the number of people living with inadequate water supplies also increases. This is called *water stress*. Stressed regions or countries are defined[13] as those where the available water is less than 1,700 cubic metres per person per year, to cover all purposes including agriculture. Between 1,000 and 1,700 cubic metres is called a 'moderate water shortage', 500–1,000 cubic metres is 'chronic' and less than 500 is 'extreme'. It is extraordinary that in 2010, 3.6 billion people, out of a world population of 6.9 billion, lived under some level of water stress – more than half the world population. The two regions which have the greatest fraction of population living with *extreme* water shortage are North Africa, with 94 million out of a total population of 209 million, and the Middle East, with 71 million out of 214 million. These figures are very high and, with these two regions increasing in population very fast, will be much worse by 2050 (North Africa 216 million out of 329 million; the Middle East 190 million out of 379 million). Per capita water availability is clearly strongly related to population rise, but it is also a function of climate change, with warming generally (but not always) leading to drying. A host of other factors also intervenes, due to the cavalier way in which we cut down trees and destroy watersheds.

There is one direct way in which ice, or the lack of it, contributes to water stress. In some parts of the world, such as northern India, the Bolivian plateau and Tibet, the water supply comes from the spring melt of snow in nearby high mountains, accompanied by run-off from glaciers. If this regular water supply fails because there is not enough snow or ice present in the first place, water stress results. There is no direct connection to sea ice here, simply global warming, but water and food shortages certainly tend to go together as two of the dark horsemen that threaten us.

II

The Secret Life of Chimneys

The story of Greenland Sea chimneys and their role in climate change is a scientifically beautiful tale that involves changes in the atmosphere, the ocean and the ice, all interacting with one another to yield a major change in the distribution of temperature over our planet. It is one that affects my own country directly.

First, what is a chimney? It is a deep, rotating, vertical cylinder of water, transporting cold water from the surface of the ocean to great depths, as much as 2,500 metres. This contradicts all that we know and hold dear about the stability of the ocean, which we think of as composed of different water masses which form horizontal layers, separated by vertical differences in salinity and temperature which give them different densities. The ocean as a whole is stable: low-density water types sit on top of water types of greater density, all the way to the bottom. Surely water at the surface can never be made to sink a mile and a half and upset this stability? Well, it can, but only in a very few vital places. One of these is the Greenland Sea. And although such features ought to be very unstable, in fact they can last for years. No one knows why.

THE GLOBAL THERMOHALINE
CIRCULATION

Let us begin with the thermohaline circulation. This is the slow churning circulation of the waters of the entire global ocean, driven not by the wind like most ocean currents, but by density variations caused by temperature and salinity differences. The main driver is in

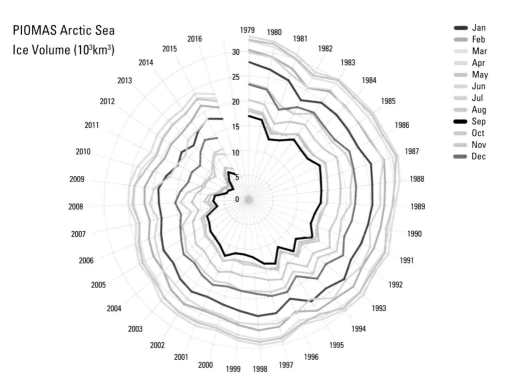

16. The 'Arctic death spiral'. Ice volume for each month of each year since 1979 on a polar plot so that a declining ice volume is seen as a spiral moving towards the centre of the graph.

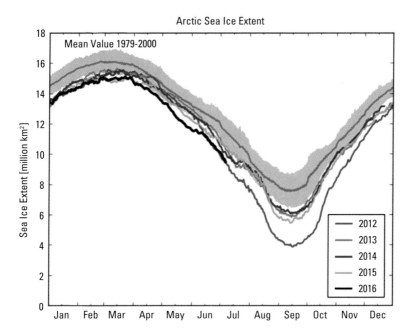

17. The seasonal cycle of sea ice extent in the Arctic. The grey band with its central median line represents the range of variation of ice extent during the era 1979–2000, before retreat became more rapid.

18. Melt pools in summer sea ice, some of which have melted right through to form thaw holes, in the Arctic Ocean.

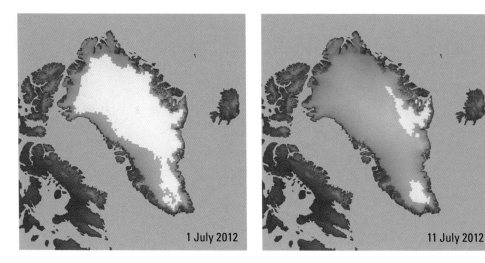

1 July 2012

11 July 2012

19. Extreme melt event in July 2012 on top of the Greenland ice sheet. The whole of the area shaded blue was detected by satellite to be wet.

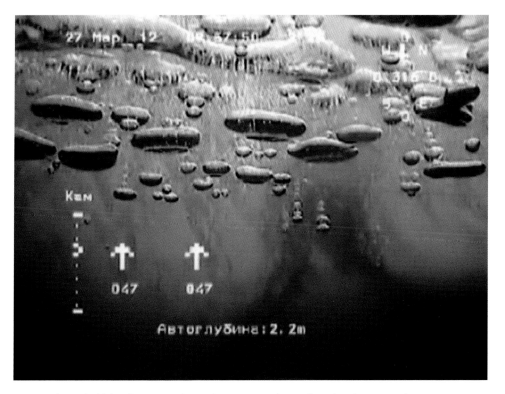

20. Methane bubbles flattening themselves against the underside of sea ice. The sea ice, seen faintly in the background, is 2.2 m thick.

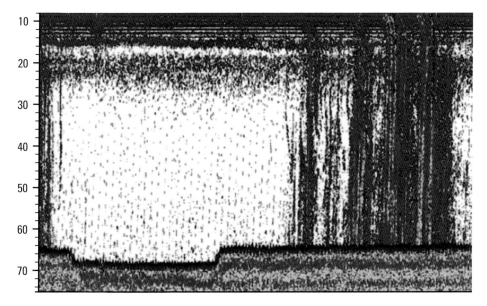

21. Rising bubble plumes on the East Siberian Shelf in water depth of 70 m, detected by sonar.

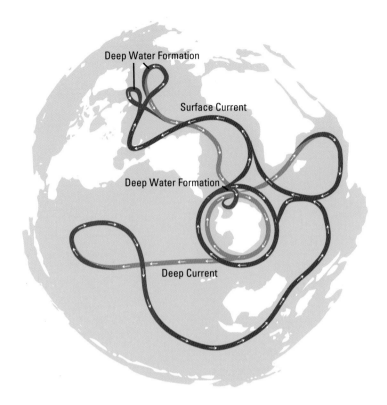

22. The global thermohaline circulation, or 'global conveyor belt', showing surface and deep components of the flow and locations of deep water formation.

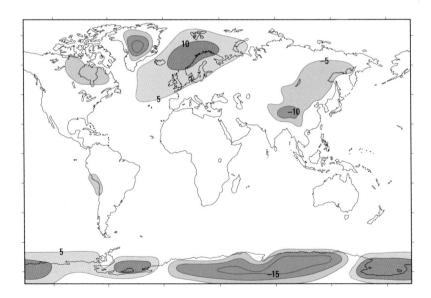

23. Temperature anomaly contours for the world, showing air temperatures relative to the zonal mean (the mean temperature for that latitude), 1999. The warm anomaly in northern and western Europe is due to warm water transported by the Gulf Stream and the Atlantic part of the thermohaline circulation.

24. Pancake ice in the Odden ice tongue, central Greenland Sea. Thick, old pancakes are being sampled from a ship.

25. Thinner, younger pancakes are being studied with a wave buoy.

26. The Odden ice tongue in winter, 1997. Red colours are heavy polar ice coming out of the Arctic Ocean. Blue and yellow colours are younger locally grown ice, which in Odden occurs as pancake ice (*inset*).

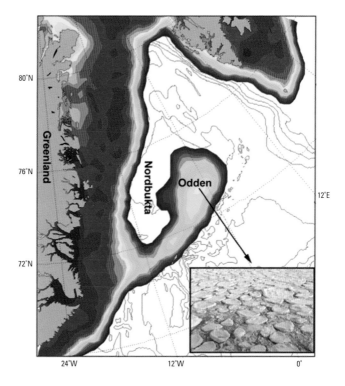

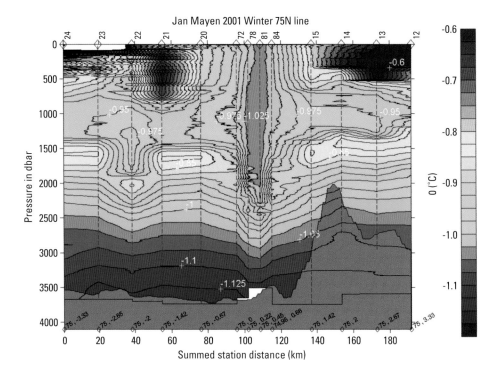

27. A temperature section through the chimney in Plate 24, also showing a second, smaller chimney nearby to the left (the pressure in decibars is almost identical to the depth in metres).

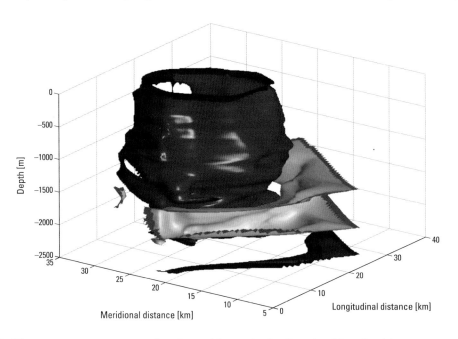

28. The temperature structure of a winter chimney in the Greenland Sea. Its chimneypot shape is picked out by the –1°C temperature contour. Note the great depth to which it sinks (2,500 m) and its perfect cylindrical structure. (It cuts through a slightly warmer layer which is at –0.9°C, shown in yellow.)

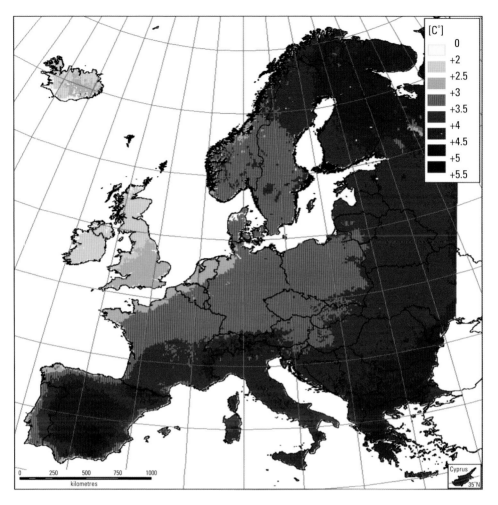

29. Projections made in 2008 by the European Environment Agency of warming of Europe by the year 2100.

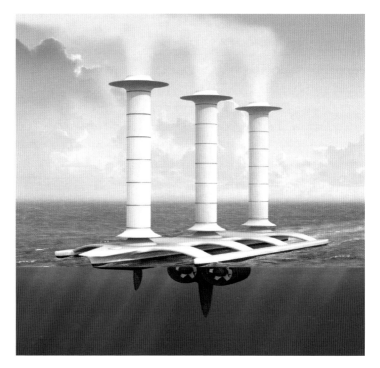

30. Marine cloud brightening spray vessel, as envisaged by Stephen Salter. Three Flettner rotors provide propulsion and are the bases for the water-particle injection into the cloud.

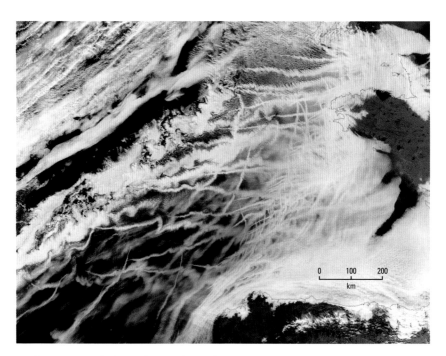

31. Condensation trails from ship tracks in the eastern North Atlantic, 44°–50°N, 5°–15°W. These are believed to survive for several days, indicating that the albedo of some maritime clouds can be enhanced by seeding from ships.

the polar oceans, where sea water at the surface gains salinity from sea ice formation (since most of the salt is rejected back into the sea, as described in Chapter 2) and sinks. This sucks in a slow flow of water from the tropics towards the poles, carrying heat and salt. The flow is modified by the shape of the continents and by the *Coriolis force* associated with the rotating Earth, which turns the moving waters to their right in the northern hemisphere and to their left in the south.

If we look at the surface circulation (Plate 22) we see a slow flow (red) which mimics the wind-driven circulation in some respects, except that it would keep happening even if there were no winds. Start by looking at the two broad areas of upwelling in the northern Indian Ocean and North Pacific. These bring deep water (blue) up to the surface which then flows slowly to the south and west. The Pacific water flows through the East Indies and joins the Indian Ocean water in flowing around the Cape of Good Hope and northwards to the tropical Atlantic. Here the flow gathers more water and heat from the Gulf of Mexico and moves northeastward across the North Atlantic, just like the Gulf Stream and North Atlantic Current. The flow continues northward towards the Arctic Ocean but then it vanishes. Somewhere the water sinks. We can then detect a slow deep flow of water back southward through the North Atlantic and South Atlantic to complete a great conveyor belt when it again reaches the Indian Ocean and Pacific – a journey which has taken it about 1,000 years. This flow was christened by Wally Broecker of Lamont-Doherty Earth Observatory the Great Ocean Conveyor Belt, which is a good name except that other scientists, such as Carl Wunsch of Woods Hole Oceanographic Institution, would say that the flow is far more complex than this and is broken up into a series of cells. But Plate 22 gives the basic feel, an inexorable slow flow of water that is driven by heating, cooling, evaporation and the rotation of the earth – very fundamental driving forces.

A conveyor belt needs to be driven. The cogs that drive the Great Ocean Conveyor are the upwelling forces which bring deep water to the surface in the Indian Ocean and Pacific, and the downwelling forces that cause surface water to sink. Because we are focusing on changes in the polar regions we will ignore the broad upwelling and examine the cogs which drive the more concentrated sinking which happens in the northern Atlantic. Where does it happen and why?

Scientists discovered that it happens only in two locations, and they are surprisingly small in extent.[1] The first is a small area in the centre of the Labrador Sea, where the surface water is chilled in winter by the cold winds blowing off Labrador and Greenland. The cooling water gains in density through the winter and eventually becomes dense enough to sink to great depths. The volume of sinking water depends crucially on the air temperatures during the winter, and varies greatly from year to year. The second area is more interesting because sea ice is involved in the process. This is a tiny area in the centre of the Greenland Sea, at 75°N 0°W, and we are going to focus on this critical site because changes there are affecting the entire world.

THE GREENLAND SEA CONVECTION SITE

The Greenland Sea is an ocean region of unique importance, situated close to Europe and intimately linked to the European climate. It is thanks to the heat transported by ocean currents up into the Greenland Sea that western Europe is 5–10°C warmer than the average temperature for its latitude (Plate 23). Were this heat transport to fail, Britain and western Europe would have a climate like Labrador.

The centre of the Greenland Sea is a window to the deep ocean. The region of sinking in the Greenland Sea occupies less than a thousandth of the area of the world ocean, yet it is vital to ocean circulation, since it is only by this sinking (also called 'ventilation') that a complete vertical as well as horizontal circulation of the ocean can take place, so that gases and nutrients dissolved in the surface waters can be cycled back into the depths. Dissolved carbon dioxide is also carried down into the depths through this sinking, which has a big impact on the ability of the ocean to absorb a significant fraction of the extra carbon dioxide that we emit into the atmosphere each year. It has been suggested that past changes in convection were responsible for some of the rapid climatic fluctuations detected in sediment and ice cores, and later in this chapter we will see that some climate models, in predicting a decline in Greenland Sea convection, also predict a consequential cooling of the western European climate.

In the Greenland Sea the convection occurs at the centre of a cyclonic gyre (that is, a rotating circulation which is anticlockwise in the northern hemisphere), bounded to the west by a cold current (the East Greenland Current, EGC) which brings polar ice and water into the system from the Arctic Basin; to the east by a warm northward-flowing current (the West Spitsbergen Current), which is an extension of the Gulf Stream; and to the south by the Jan Mayen Current, a cold current offshoot of the EGC which is diverted to the eastward from the main EGC at about 72–73°N by the presence of a subsea mountain chain, the Jan Mayen Ridge (Plate 26). The Greenland Sea also represents the main highway for water and heat exchange between the Arctic Ocean and the rest of the world, because Fram Strait, which connects the Greenland Sea to the Arctic Ocean, is the only deep water entrance to the Arctic. Ice is transported down into the Greenland Sea from the Arctic Ocean and melts as it moves southward, so that the Greenland Sea as a whole, when averaged over a year, is an ice sink and thus a fresh water source. Melting ice contributes some 3,000 km^3 per year of fresh water to the Greenland Sea.

However, in winter local ice can also form within the Greenland Sea itself, within the region of cold water which has moved eastward in the Jan Mayen Current. The water is already cold because it comes from the East Greenland Current. It leaves behind its polar ice cover, which carries on southward down the coast of Greenland. This leaves the cold but ice-free water exposed to further intense cooling from a cold atmosphere in winter, particularly during phases of the climate when the prevailing winds in winter are westerly, blowing off the Greenland ice cap. The intense cooling causes new sea ice to form on this cold open water. But the new ice cannot form a continuous sheet, because of the huge amount of wave energy found in the Greenland Sea in winter. Instead it follows the classic 'frazil-pancake cycle', forming initially as a milk-of-magnesia suspension of frazil ice crystals in the water column, then as small cakes only 1–5 metres across, which acquire raised edges from their frequent collisions (Plates 24–26). The cakes are formed by waves causing the crystals in the frazil suspension to squeeze together into clumps. These pancakes, and the frazil that they float in, completely fill the area of sea surface which carries the cold polar water of the Jan Mayen Current. This can be seen in satellite images (Plate 26), and the new ice forms a tongue-shaped protrusion called the Odden Ice Tongue, which can occupy an area of up to 250,000 km². This was first discovered and named in the nineteenth century by sealers, because harp seals use the small cakes to haul out on in spring and give birth to their pups. Norwegian sealers would follow the outer ice edge of the tongue to slaughter the young pups for their valuable white fur and called the area Odden (Norwegian for headland). The area was known to early whalers as well, since the partially protected bay of open water to the west of the Odden, Nordbukta ('Northern Bight'), was an area frequented by the slow-swimming right whale (or bowhead whale). The great whaler-scientist William Scoresby Jr, whom we met in Chapter 6, wrote about this ice tongue and bay in his classic 1820 book, *An Account of the Arctic Regions With a History and Description of the Greenland Whale-Fishery.*[2]

The exciting thing about pancake ice formation is that most of the salt contained in the water that freezes does not enter the structure of the ice but is rejected back into the ocean. In experiments by my research group, we lifted pancakes by crane on to the deck of a ship and sliced them up. We found that thinner pancakes have a salinity of about 10 parts per thousand (compared to 35 for ocean water), while the thicker ones can have a salinity as low as 4 ppt, having lost nearly 90 per cent of their salt. The brine release increases the density of the surface water and adds to the effect of cooling to destabilize the surface layer and cause the surface water to sink.[3] It is a much more powerful effect than in the Labrador Sea because of the extra density impact of the salt. In fact, the rapid growth rate of pancake ice, and thus the rapid increase in the brine content of the surface water, is crucial to the large amount of convection occurring in the Greenland Sea and thus to the maintenance of the Atlantic thermohaline circulation. This is what makes pancake salt rejection exciting; it is a rapid process since pancake growth is rapid, and it happens in just the spot where it can have a big impact on ocean stability.

Because of the importance of the Odden to Scandinavian sealers, its extent has been recorded nearly every year since 1855, even before the Danish Meteorological Institute was founded and started publishing monthly ice reports. It used to form almost every winter in November, lasting until April or even May, so we can assume that convection took place throughout that period. But since the 1990s something has happened to disturb this. In 1994–5, and from 1998 to the present, Odden has failed to develop altogether. It is a major change in the nature of the Greenland Sea. Why did this happen? Partly because the climate switched into a new phase where the prevailing winds over the Odden area came from the east and were warmer (this switch between two atmospheric circulation systems is called the North Atlantic Oscillation or NAO). But, more seriously, when the NAO switched back into its former phase, the Odden did not come back because global warming had made air temperatures over the sea warm enough to prevent its formation.

THE SECRET OF THE CHIMNEYS

What does this mean? How has it affected convection, the sinking of the surface water to great depths? To be able to judge, we have to examine how convection happens; and it turns out that there is another wonderful process, parts of which are not understood, called *chimney* formation. Chimneys were first discovered in a warm part of the ocean, the Gulf of Lion in the northwest Mediterranean, in 1970, during a big oceanographic experiment called MEDOC (Mediterranean Ocean Circulation Experiment).[4] It was found that at times in winter when the mistral, an intensely cold northwesterly wind, blew out to sea from the Alpes Maritimes, the cold air chilled the surface water to the extent that the water sank, not in random masses but in the form of small coherent rotating cylinders, which were named chimneys. A chimney did not last long – only a few days – because the winds would change direction, and they were regarded as an amusing curiosity. Then in the 1990s scientists working in the Greenland Sea began to suspect that this was how convection under the Odden occurred. I had been leading a European Union project called the European Sub Ocean Programme (ESOP), which then metamorphosed into a second programme, called

Convection, which was also funded by the EU. Thanks to this funding and the participation of many oceanographic institutions that possessed ships, such as the Alfred Wegener Institute in Bremerhaven and the Norwegian Polar Institute in Tromsø, it was possible to run voyages into the centre of the Odden region during midwinter to see how convection took place.

What we all found was remarkable (Plates 27 and 28). Over a tiny distance, only 20 kilometres across, the surface water forms a tight cylinder, rotating like a solid body in a clockwise direction (opposite to the Greenland Sea gyre as a whole), carrying water downwards and extending its influence to the staggering depth of 2,500 metres, in an ocean which is only 3,500 metres deep.[5] The surface water, made enormously denser by ice formation as well as cooling, has to sink that far before it encounters surrounding water which is equally dense and so stops it from sinking further. In sinking this distance, the tight cylinder cuts through everything in its way, including a deep layer of warmer water which the sinking water simply punches through. The shape of the chimney in Plate 28 is traced out by the −1.0°C temperature contour (suitably coloured red to look like a chimney pot), which cuts through a yellow deep layer of slightly warmer −0.9°C water. The presence of the chimney can be tracked whether we plot temperature, salinity or density. The chimney in Plate 28 has a smaller chimney nearby, which has not sunk so far. Plate 27 shows the two chimneys in a temperature cross-section, produced by running a line of oceanographic stations, in which a probe is lowered to measure temperature and salinity, right across the centres of the two features.

The amazingly coherent nature of this rotating cylinder is shown in fig. 11.1, where the Alfred Wegener Institute ship FS *Polarstern* positioned herself over a chimney and measured the velocity of the water in it using an acoustic device (an ADCP, or acoustic döppler current profiler).[6] We can see that within the cylinder the water is rotating at a speed that is proportional to its distance from the centre – in other words, like a solid rotating mass. And let's remember that this rotation, which is clockwise (called anticyclonic in oceanography), is directly opposite to the generally anticlockwise rotation of the currents

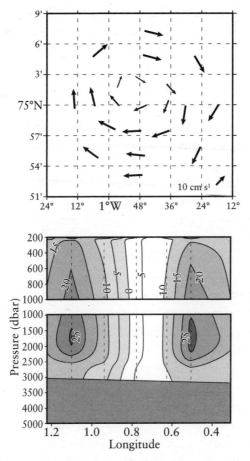

Figure 11.1: Water current velocities around the centre of a Greenland Sea chimney, showing solid body rotation. The top diagram is a surface view looking straight down; the bottom cross-section shows that solid body rotation extends right down through the cylinder (the region on the left shows water flow out of the plane of the paper; the right flow into the plane of the paper).

in the Greenland Sea, another reason for our amazement that this cylinder can form and persist.

How many chimneys exist? Our problem with a small research ship – we used the handy Norwegian vessels *Lance* and *Jan Mayen* – was that having found a chimney it took us the rest of the cruise to map it adequately. The foul weather in the Greenland Sea in winter meant that we often had to stop work and heave to – on one memorable occasion the eye of a storm passed right across us, dropping the air pressure to only 917 millibars, during a deceptively peaceful hour before the second half of the Force 12 storm hurled itself at us. The largest number of chimneys that we found in a single survey was two (most winters we found only one, which obligingly sat at exactly 75°N 0°W), and that survey was carried out in a miraculously quiet winter period when we felt that our stations were close enough to detect any chimney that was present.[7] We therefore suspect that there were indeed only two chimneys in the central Greenland Sea that year. Re-analysis of studies made in earlier years suggests that there used to be many more chimneys: Jean-Claude Gascard, from the Université Pierre et Marie Curie in Paris, deployed a series of neutrally buoyant floats (which are weighted to float at pre-arranged depths) in 1997, and found that at any one time four of them would be turning in tight circles at depths of 240–530 metres, which we later realized must have meant that they were trapped in chimneys. So in the 1990s there were many more chimneys than in the 2000s. It is no accident that there was more ice too.

During the Convection project we visited the centre of the gyre for three winters running (2001–3), while our colleagues in the Alfred Wegener Institute visited during the intervening summers. We found some extraordinary things. A chimney is very long lived. An open chimney found in the first winter was found to still be in the same position in the subsequent summer, but was capped by 50 metres of the fresher, less dense water which covers the surface of the summer Greenland Sea because of melt from sea ice and glaciers. Underneath this cap the chimney continued to exist as a rotating submerged cell. The following winter it had opened itself up to the surface again, and returned to functioning as a convective centre. The process was repeated in the subsequent summer and winter until the end of the project meant that we could not follow it further. This is the longest-lived ocean chimney ever studied.[8] Such longevity in such a small, tight feature is unknown elsewhere in the ocean, where features the size of eddies lose energy and momentum by friction and 'run down' after a few days or weeks. We don't know what maintains a chimney in such a rapidly rotating, compressed state. Why does it not run down? Nor do we know why a chimney stays so exactly in one place – our longest-lived chimney moved only 10 km during three years, despite there being no feature on the seabed to anchor it to one location, such as often happens with ocean eddies. Chimneys remain a mystery in many ways. We were profoundly disappointed that, having made these key discoveries with huge climatic implications, our repeated bids to the Natural Environment Research Council (NERC) in the UK for further support to study chimneys in the field were all turned down.

What we do know is that there are now fewer of these remarkable structures, coinciding with a loss of ice from Odden, and that this decrease in convection in the Greenland Sea will have a serious impact on the climate of the whole world ocean. Models suggest that between six and twelve chimneys need to form and dissipate per year to account for the amount of deep water formation going on. Where are these chimneys now? Do they still exist despite the difficulty of forming without ice? Is deep water formation slowing or stopping? Or is it happening in a different way, or in a different place?

THE GREAT CONVEYOR BELT ON THE DAY AFTER TOMORROW

In 2004 a dramatically inaccurate film was released called *The Day After Tomorrow*. In it, a decrease in convection due to meltwater covering the polar seas leads to a change of climate in which New York is turned into an icy polar desert within a period of days. Better not mess with the thermohaline circulation was the message. But we are messing with it. There is already evidence that the decline in convection in high latitudes is leading to a reduction in the transport of heat in the Atlantic. The overall strength of the Atlantic thermohaline circulation is estimated to be 15–20 Sverdrups (millions of tons of water per second), transporting 1 petawatt (a thousand million million watts) of heat northward. Deep currents flowing southward past the Faroes have been observed to decline in strength, showing that one part of the conveyor belt, the part that carries water away from the Arctic at depth, has weakened. A weakening of the thermohaline circulation at the surface of the North Atlantic may not be noticed for a while, because the flow is dominated by the wind-driven Gulf Stream and North Atlantic Current. But a persistent loss of current in this warm part of the conveyor belt will eventually be noticed.

Will it chill our climate, at least the climate of those of us that live on the Atlantic seaboard of Europe? Yes, it will, though not as rapidly or severely as is made out in the film. In fact all that it will probably mean is that our region warms up more slowly than continental Europe. Plate 29 shows the projections of a climate model run by the European Environment Agency in which a standard 'business as usual' scenario gives us an expected, and feared, warming of 4°C over most of Europe by 2100, a warming which will turn the climate of southern Europe into something resembling that of North Africa today. But in this case the model simulation included a decrease in the Atlantic thermohaline circulation, and this greatly reduced the warming that will be experienced by Britain, Ireland, Iceland and the Atlantic coastlines of Norway, the Low Countries and France. In fact Britain's warming is halved to 2°C. In retribution, of course, the retention of more heat at low latitudes will mean a faster warming of the surface water of the tropical Atlantic, probably leading to hurricanes of much greater intensity.

Possibly connected with the loss of convection in the Greenland Sea is the fact that a new convection site was discovered in 2003, in the Irminger Sea just to the east of the southern tip of Greenland.[9] Convection is much shallower than in the Greenland Sea, reaching only to 400 metres in most winters and 1,000 metres in exceptionally cold winters. The form of the convection is quite different – no neat, mysterious cylinders, just a diffuse mass of sunken water occupying a relatively large area, which moves on to the southwest afterwards to interact with Labrador Sea water. It seems to be a kind of precursor of the Labrador Sea convection, and ice is not involved in the process.

THE FUTURE

The immediate impacts of Greenland Sea research are already clear. Climate models will have to be adjusted to take account of the mechanisms discovered by the Convection project. But we have to remember that Convection was simply a physics project seeking to understand the physical behaviour of a part of the ocean which is also a habitat for life. Biology and chemistry need to take account of the results of the project too. One discovery, made by Jan Backhaus at the University of Hamburg, is that the quantity of plankton inside a chimney in winter, per unit area of sea surface, is as great as that observed in spring and summer. The reason is that the whole 2,500-metre column of water has the same density, so plankton can rise towards the surface and sink for great distances, all without effort, in search of nourishment. Even though it is dark in winter, a convecting chimney can support more life than a similar area of normal ocean.

The challenges that remain are to follow the development of the central Greenland Sea as present structures break up and new ones develop, and to predict the implications for the volumes and depths of convected waters as climate, and thus surface forcing, changes. As in so many areas of climate science, further research is urgently needed to avoid nasty surprises, and as in so many areas, such research is not being funded. This is another aspect of the mental self-censorship that seems to occur among some climate scientists and most science funding bodies. Numerous proposals for work in the central Greenland Sea have been turned down in recent years, not just by the British. Yet everyone accepts that the thermohaline circulation is a vital part of our climate system and its change or disruption would have major global effects. So it is vital to send scientists back out there.

12

What's Happening to the Antarctic?

THE STRANGE STORY OF ANTARCTIC SEA ICE

My focus in this book so far has been on the Arctic. This is for very good reasons: the Arctic is the backyard of most of the advanced industrial countries of Eurasia and North America, and the rapid changes occurring there affect us immediately. For Britain, sea ice in the Greenland Sea begins only 400 miles from the Shetland Islands. For Canada, Russia and the USA, sea ice forms part of their territorial waters. By contrast, the Antarctic is far away; it is even far away from any land masses. At its narrowest point, the Drake Passage, which separates Antarctica from South America, is still 1,200 miles wide. Does it matter what happens down there?

It matters for at least two reasons. One is that the albedo feedback effect from snow and ice retreat must be computed for the planet as a whole. We have discussed in previous chapters how the Arctic sea ice is retreating very rapidly. We know that the speed of this retreat greatly exceeds the projections of most climate models, which predict a slower retreat in tune with general global warming. In this respect Arctic sea ice is an anomaly. But Antarctic sea ice is an even greater anomaly, because it is actually advancing. Not very much, but it is advancing, despite an overall warming over the Antarctic continent. If Antarctic sea ice is advancing then this will help to offset the global albedo reduction due to Arctic sea ice and snowline retreat. Secondly, the advance of Antarctic sea ice is just as much of a challenge to global climatic models as the rapid retreat of Arctic sea ice – computer models predict a slow retreat in both polar regions, so they are wrong at both ends. In September 2013 Antarctic ice extent reached a record maximum of 19.47 million km², according to the US National Snow and Ice Data Center (NSIDC) in Boulder. This is approximately 30,000 km² larger than the previous record set in 2012, and is 2.6 per cent higher than the 1981–2010 average. In more recent years the area has fallen back somewhat, to 18.83 million km² in 2015, possibly associated with the onset of an El Niño atmospheric pattern in the southern hemisphere,[1] but it still shows a slow increasing trend.

We know from passive microwave instruments on satellites that Antarctic sea ice is advancing overall. This is occurring despite the fact that at least one part of the Antarctic – the Antarctic Peninsula – is warming very fast,[2] and that this led in 2002 to a spectacular event – the collapse of the Larsen B ice shelf on the east side of the Peninsula – in which an area of 3,250 km² of ice shelf, 200 metres or more thick, broke up into an army of icebergs which then proceeded to drift away, leaving islands and coastline accessible to shipping for the first time in recorded history.

Why, then, is Antarctic sea ice advancing in the face of a warming climate and the loss of ice shelf area? To answer that question we need to understand how Antarctic sea ice differs from Arctic ice. Of course, Antarctica differs from the Arctic in that the Arctic is made up of ocean surrounded by land, while the Antarctic is a huge land mass at the Pole surrounded by a vast ocean. (Interestingly, the size and shape of the Arctic Ocean and the Antarctic continent are very similar.) Wind patterns and ocean currents tend to isolate Antarctica from global weather patterns, so it can exhibit trends which are uncoupled from those of the Arctic.

WHY ANTARCTIC ICE IS DIFFERENT

Antarctic sea ice is not the same as Arctic sea ice. Of course they are both made of frozen water, but Antarctic sea ice forms in a different way, and has different properties and appearance from Arctic ice. New sea ice starts to form early in the winter close to the Antarctic coast, and the ice edge advances northward into the great Southern

Ocean as winter progresses, exposed to all the power of the world's biggest ocean. The mechanism by which the sea ice is generated was not understood until an expedition was able to work in the pack ice zone during early winter, the time of ice edge advance. This occurred for the first time in 1986, in the Winter Weddell Sea Project, using the German research ship FS *Polarstern*. I was aboard that ship on her memorable cruise, along with fifty other scientists. As we traversed the ice margin region we made a careful study of the ice conditions and characteristics, and identified what we called the *frazil-pancake cycle* as the source of most of the first-year sea ice seen further inside the pack.[3]

We have seen (Chapter 2) that ice growing on calm water forms an initial skim which solidifies into a thin transparent layer called nilas; water molecules freezing on to the bottom of a nilas sheet then extend the ice downwards, with a selection factor favouring crystals with horizontal c-axes, to yield eventually a first-year ice sheet. Ice forming at the extreme Antarctic ice edge cannot grow directly into a continuous sheet of nilas like this because of the high energy and turbulence in the Southern Ocean wave field, which maintains the new ice as a dense suspension of frazil ice. This suspension undergoes cyclic compression because of the particle orbits in the wave field, and during the compression phase the crystals can freeze together to form small coherent cakes of slush, which grow larger by accretion from the frazil ice and more solid through continued freezing between the crystals. This becomes known as *pancake ice* because collisions between the cakes pump frazil ice suspension on to the edges of the cakes, then the water drains away to leave a raised rim of frazil ice which gives each cake the appearance of a pancake. At the ice edge the pancakes are only a few centimetres in diameter, but they gradually grow in diameter and thickness with increasing distance from the ice edge: they may reach 3–5 m in diameter and 50–70 cm in thickness. The surrounding frazil continues to grow and supply material to the growing pancakes, since the water surface is not completely closed off by ice and a large ocean–atmosphere heat flux is still possible which can dispose of latent heat. This is exactly the same mechanism that creates the Odden ice tongue of the Greenland Sea, as described in Chapter 11, but differences occur as the ice develops further.

At greater distances inside the ice edge, and now acquiring some protection from waves because of wave energy loss at the edge, the pancakes begin to freeze together in groups; but during our winter experiment in 1986 the wave field was found to be powerful enough to prevent overall freezing until a penetration of some 270 km was reached. Here the pancakes coalesced to form first large floes, then finally a continuous sheet of first-year ice. At this point, with the open water surface cut off, the growth rate dropped to a very low level (estimated at 0.4 cm per day[4]) and the ultimate thickness reached by first-year ice was only a few centimetres more than had been attained when the pancakes consolidated.[5]

First-year ice formed in this way is known as *consolidated pancake ice* and has a different bottom shape from Arctic ice. The pancakes at the time of consolidation are jumbled together and rafted over one another, and they freeze together in this configuration with the frazil acting as 'glue'. The result is a very rough, jagged bottom shape, with rafted cakes doubling or tripling the normal ice thickness, and with the edges of pancakes protruding from the upper ice surface to give a landscape which I described in a paper as a 'stony field' because of its resemblance to a landscape of tiny fields surrounded by drystone walls. The contrast between such ice and ice formed in calm conditions is shown in fig. 12.1, which depicts profiles generated by drilling holes at 1-metre intervals. Hole drilling, though laborious, is our best way of mapping the shape of the ice underside, since the Antarctic Treaty of 1959 does not allow submarines to work in Antarctic waters.

The rafted bottom of consolidated pancake ice provides a large surface area per unit area of sea surface, providing an excellent substrate for algal growth and a refuge for krill. The thin ice permits plenty of light to penetrate, allowing phytoplankton to photosynthesize and live on the underside of the ice. The result is a fertile winter ice ecosystem, which is thought to contribute about 30 per cent of the total biological production of the Southern Ocean.

Even after another thirty years not many ships have worked in the Antarctic pack ice in midwinter. The Alfred Wegener Institute ran a Winter Weddell Gyre Study in 1989,[6] in which I participated, and there were more recent experiments in the Weddell Sea where the ship worked in conjunction with an ice camp – Ice Station Weddell-1 in

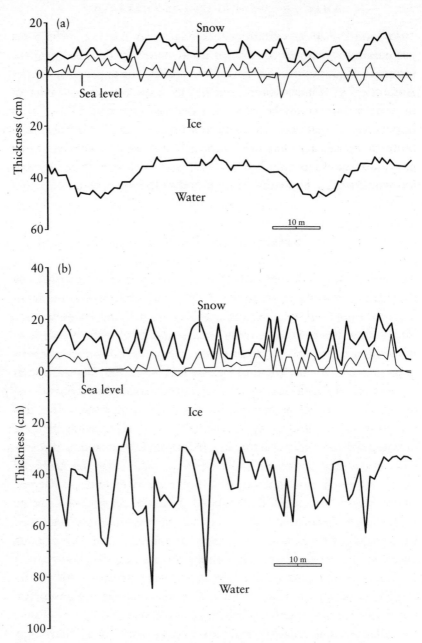

Figure 12.1: Winter thickness profiles across Antarctic ice fields, from holes drilled at 1-metre intervals, showing the difference between ice which has grown in calm conditions with a smooth underside (a) and consolidated pancake ice, where the pancakes have frozen together in a jumbled fashion to give a jagged underside with rafting to two or three pancake thicknesses (b).

1992[7] and ISPOL (Ice Station POLarstern) in 2004–5.[8] We do not yet have enough evidence to be sure of whether the frazil-pancake sequence of ice growth is followed around the entire periphery of the Antarctic, but, if it is, then the area occupied by Antarctic pancake ice in early winter could be as great as 6 million km^2, making it an important yet seldom-seen component of the Earth's surface. It is quite extraordinary that this amazing landscape of heaving white pancakes should occupy such a vast area yet should be so little known. Probably fewer than a thousand people have seen it.

SNOW ON THE ICE

The annual snowfall on to Antarctic sea ice is much greater than in the Arctic because the proximity of the vast Southern Ocean brings more moisture and so more precipitation, and in coastal regions snow is also blown on to the sea ice by katabatic winds (winds which blow down the slope of the Antarctic ice sheet from its summit) off the tops of ice shelves. During the July–September 1986 *Polarstern* cruise in the eastern Weddell Sea we found a mean snow thickness of 14–16 cm on the surface of first-year ice. Since the ice itself is so thin, this was sufficient to push the ice surface below sea level in 15–20 per cent of holes drilled, leading to the infiltration of sea water into the overlying snow and the formation of either a wet slushy layer on top of the ice or, in the case of freezing, of a 'snow-ice' layer between the unwetted snow and the original ice upper surface. In September–October 1989 the snow was even thicker, especially over multi-year ice in the western Weddell Sea into which we ventured. This was sufficient to push the ice surface below sea level in almost every case. Fig. 12.2 (a) and (b) show the contrast between these two types of ice cover. The thick snow insulates the ice and its slushy wetness means that satellite radar methods for mapping ice thickness do not work well because the radar beam is reflected by the wet snow. There is no doubt that snow, and the slushy ice formed from water infiltrating into the snow (called 'meteoric ice'), play a much bigger role in Antarctic sea ice than in the Arctic.[9]

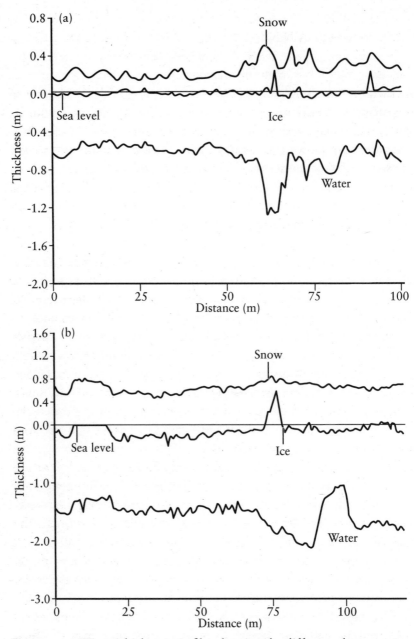

Figure 12.2: Winter thickness profiles showing the difference between first-year ice (a) and multi-year ice (b) in the western Weddell Sea, demonstrating the way in which the weight of the snow cover pushes the ice surface below sea level, especially for multi-year ice.

THE ANNUAL ICE CYCLE AND ITS CHANGES

Awkwardly for climate change models, as I said at the beginning of this chapter, the extent of Antarctic sea ice has been observed to be slowly increasing in recent years, but with a high regional variability.

Figure 12.3 shows the annual cycle of ice extent for the period 1978–2011.[10] In summer there are only two substantial areas of sea ice remaining, in the western Weddell and Ross seas, so these are the only regions that can contain much multi-year ice, the ice type which until recently was dominant in the Arctic. The minimum varies little from year to year. As winter sets in, new ice forms north of the ice edge and the sea ice limit advances northward until it reaches a maximum at between 55°S and 66°S by the end of winter (August–September), whereupon it retreats back to its start point. The northern limit is 55°S in the Indian Ocean sector at about 15°E, but lies at about 60°S around most of the rest of East Antarctica, then slips even further south to 65°S off the Ross Sea. The edge moves slightly north to 62°S at 150°W, then again shifts southward to 66°S off the Amundsen Sea before finally moving north to engulf the South Shetland and South Orkney Islands off the Antarctic Peninsula and complete the circle. The variation in latitude of this winter maximum as we move around the Antarctic therefore amounts to some 11°.

The absolute limit of northward ice advance is the edge of the Antarctic Circumpolar Current, where surface water temperature changes abruptly in the Polar Front, or Antarctic Convergence. Everything else changes here as well – on heading south in a ship it is the point at which you start to encounter icebergs, penguins, albatrosses, skuas and other Antarctic birds in profusion, as well as rich plankton (the famous shrimp-like krill) and the great whales which eat them. The sea turns green and there is a smell of life in the air. The ice seldom reaches this natural ocean boundary, however, because its advance is limited both by ocean processes (storms, eddies) which break it up, and by surface air temperatures: it was shown by Jay Zwally of NASA and his colleagues[11] that the winter advance of the ice edge follows closely the advance of surface air which is colder

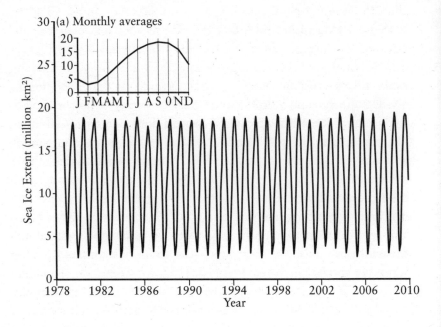

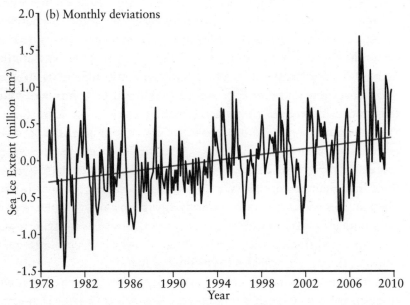

Figure 12.3: (a) Monthly averaged southern hemisphere sea ice extents for November 1978 to December 2011. The inset shows the average annual cycle. (b) Monthly deviations for the sea ice extents.

than the freezing point of sea water (−1.8°C), and almost coincides with this temperature line (or isotherm) at the time of maximum advance. The magnitude of this annual cycle in extent (defined as the area south of the main ice edge) can be easily measured by satellites, especially the useful NASA passive microwave satellites (called SMMR, SSM/I and SSMIS), and fig. 12.3 shows results obtained by the group at the NASA Goddard Space Flight Center in Greenbelt, Maryland, over the period 1978 to 2011.[12] The average maximum and minimum ice extents for the period were 18.5 million km² and 3.1 million km².

As fig. 12.3 shows, there is clearly a slow upward trend in the winter maximum extent for the Antarctic as a whole, amounting to 17,100 km² per year. The trend, however, conceals substantial regional and seasonal variations. The most rapid growth has been in the Ross Sea sector (13,700 km² per year), with lesser contributions from the Indian Ocean sector and eastern Weddell Sea, while the Bellingshausen/ Amundsen seas of West Antarctica have experienced a retreat rate of 8,200 km² per year. Eric Steig of the University of Washington found that air temperatures over the Pacific sector of the Antarctic continent (Antarctic Peninsula to the Ross Sea) have warmed twice as fast as over the rest of the continent,[13] while an analysis of the temperature record at Byrd Station (120°W longitude) shows that from 1958 to 2010 it warmed by between 1.6 and 3.2°C, a very large increase.[14] The fast warming over the Pacific sector of Antarctica (West Antarctica) is reflected in a decrease in the length of the ice-covered season (number of days per year in which a given location has an ice cover) by one to three days per year between 1979 and 2010,[15] while the Atlantic–Indian Ocean sectors showed a slower increase in the ice-covered season. The message from the ice cover is clear: a wide swathe of East Antarctica has an ice cover which is growing slowly, while a narrower swathe of West Antarctica has an ice cover which is shrinking more rapidly, the net effect being a very slow growth.

There are other, more detailed, ice variations related to local topo-graphic factors, and usually visible in spring or summer. In the sector off Enderby Land at 0°-20°E a large gulf opens up in December to join a coastal region of reduced ice concentration which opens in November. This is a much attenuated version of a mysterious winter polynya which was detected in the middle of the pack ice in this sector during 1974–6[16] but which has not been seen, at least as a full open water feature, since that date. It was known as the Weddell Polynya and lay over the Maud Rise, a plateau of reduced water depth. The area was investigated in winter 1986 by FS *Polarstern*, and it was found that the region was part of an area called the Antarctic Divergence, where upwelling of warmer deep water can occur, allowing enough heat to reach the surface to keep the region ice-free in winter.[17] As this has not occurred since 1976, the region is presumably balanced on the edge of instability as far as its winter ice cover is concerned. The 1986 winter cover was of high concentration but very thin.[18] The December distribution also shows a recurrent open water region appearing in the Ross Sea, the so-called Ross Sea Polynya, with ice still present to the north of it. In November and December a series of small coastal polynyas can be seen actively opening along the East Antarctica coast, mostly driven by offshore (katabatic) winds driving ice away from the coast as fast as it can form.

WHAT IS HAPPENING TO THE ICE?

Much of the ice in the winter pack is of pancake origin and is quite thin. Given that the climate of much of the Antarctic is warming, why is the ice limit advancing rather than retreating, with the regional pattern described above?

A simple explanation of the *circumpolar* expansion of the ice (i.e. the ice over the whole Antarctic) came from Jinlun Zhang of the University of Washington. He suggested that it is the result of strengthening winds around the Antarctic continent.[19] The key is the great circumpolar west wind belt, also called the polar vortex. Satellites have been able to measure the strength of these winds since the 1970s, and they have been growing steadily more powerful. The average wind speed is higher, and the wind is coming mainly from the west. Consider a typical ice floe. It is blown eastward by the direct force of the wind stress on its surface, but as it moves it experiences a force tending to turn it to the left, that is, to give it some component of motion northward. This is the well-known Coriolis force due to the rotation of the Earth: it acts to the right in the northern hemisphere, to the left in the southern, and is zero at the Equator. It acts because we make all our measurements of moving objects relative to a frame of reference fixed to the Earth's surface (for instance, N–S and E–W axes), but because the Earth rotates its surface is actually an accelerating frame of reference, so a moving object does not go in a straight line but deviates to left or right.

The Coriolis force is proportional to the speed of the object relative to the Earth's surface, so, with a higher wind speed, the northward-acting Coriolis force on the ice floe increases, moving the floe northward more rapidly, even as its main motion is eastward. Therefore, although it will reach a latitude where the warmer atmosphere will melt it, its northward speed will take it further north before this happens. So the whole Antarctic pack ice zone is like a great merry-go-round driven by the wind, which is throwing its ice northward into warmer water. This may, however, be a too simplistic view. First, the mechanism would lead to an increase in ice *extent* but not necessarily of *area*, since it deals only with the dynamics of existing ice. We can explain this by saying that the ice moving north in winter leaves open water behind it which quickly freezes under winter conditions. Secondly, the increase is bound to be temporary as eventually global warming will win out over increased wind speeds and the ice will fail to reach the lower latitudes. However, the mechanism is rooted in simple physics and the observed fact that circumpolar wind speeds have indeed increased.

My own explanation is based on the frazil-pancake cycle, described above, and its interaction with these enhanced winds. As the stronger winds blow around Antarctica they create bigger and longer ocean waves. The longer waves can penetrate further into the marginal ice zone and can maintain it as frazil-pancake ice to a greater distance from the ice edge. We know that frazil-pancake ice grows much faster than continuous ice because the atmosphere is not cut off from the water below, so ocean heat can be more easily lost to the atmosphere and permit faster ice growth. Could it just be that in an era of stronger winds and bigger waves we have a wider frazil-pancake zone, growing ice more quickly?

RESPONSE OF THE ANTARCTIC
TO CHANGES ELSEWHERE

To explain the *regional* nature of the sea ice trends we need a model which considers whether forcing from elsewhere can cause regional effects on the Antarctic ice.

One obvious cause, although the effects will be long-term, is the Antarctic ice sheet itself, which is starting to lose mass,[20] although more slowly than the Greenland ice sheet. An estimate presented at the May 2016 ESA Living Planet Conference in Prague is that net Antarctic ice loss at present is about 84 Gt per year, compared to at least 300 Gt from Greenland. If the rate of Antarctic ice loss increases, it has been projected that the Filchner-Ronne and Ross ice shelves will disintegrate, allowing Antarctic glaciers (for instance, those in the Transantarctic Mountains) to debouch directly into the ocean. This will rapidly accelerate the rate of mass loss from Antarctica, leading to acceleration in global sea level rise, but will also impact Antarctic sea ice (if it still exists by then). Such changes are not predicted to occur for a few centuries, except for the possible case of a disintegration of the ice shelf around Pine Island Bay and a region of East Antarctica where the ice sheet is believed to be potentially unstable should a 'plug' of coastal ice decay.[21]

For more immediate effects, which are determining the *present* regional variation in Antarctic sea ice advance or retreat, we need to look for teleconnections (long-distance links) which can exist with the lower latitude oceans and atmosphere and even with northern latitudes extending up to the Arctic. There are many candidates for the linking mechanism. Reg Peterson and Warren White[22] from the Scripps Institution of Oceanography considered the Antarctic Circumpolar Wave, a system of waves on the Antarctic Circumpolar Current which propagate slowly eastward (though westward relative to the current) and may interact with the tropical El Niño–Southern Oscillation (ENSO) system. The El Niño (Holy Child Current) is a warm ocean current of variable intensity that develops in late December (hence the name) along the coast of Ecuador and Peru and sometimes causes catastrophic weather conditions, but its name is now given to an entire South Pacific-wide anomaly of winds and currents. More recent work focuses on the Southern Annular Mode (SAM),[23] which is another complex mode of variability in the atmospheric circulation at high latitudes. There have been suggestions[24] that an El Niño year leads to more sea ice in the Weddell Sea and less sea ice in the Pacific, with the opposite for a La Niña year ('La Niña' refers to a cooling of the ocean surface off the western coast of South America, which occurs periodically every four to twelve years and affects the Pacific and other weather patterns; it is the opposite to an El Niño); but the ENSO link is complicated by recent discoveries about Central Pacific El Niño events.[25] Wider-ranging latitudinal teleconnections could relate to the link between Arctic warming and lower latitude weather extremes due to distortions in the jet stream,[26] which might then involve onward links with tropical and southern hemisphere patterns.

Any complete explanation for why Antarctic and Arctic sea ice behaviour differ must also depend on the fundamental differences between Arctic and Antarctic sea ice. The Antarctic is bound to warm more slowly than the Arctic because of the greater area of ocean with its high heat capacity and the way in which the Antarctic Circumpolar Current insulates the continent from the warmer ocean to the north. The Antarctic sea ice limits are set in a different way from the Arctic: in summer the ice retreats to the land, leaving substantial ice mass only in awkwardly shaped bights like the Weddell Sea, while the winter limits are thermodynamic and set by conditions in the open ocean. In the Arctic the situation is the opposite: the *winter* limit is set by surrounding land masses, while in summer the ice retreats to an ocean limit that is thermodynamically and dynamically set. Albedo feedback is also less important in the Antarctic than the Arctic, since at the time of maximum summer insolation (solar radiation) in late December the Antarctic sea ice has already retreated almost to the continent, while the Arctic sea ice at the summer maximum for solar radiation (June) still has a long time to go to retreat to its September minimum and is thus susceptible to changes in forcing.

A final point about the rate of warming is that the rapid Arctic warming itself creates feedbacks which lead to a further acceleration of warming. As well as the ice albedo feedback these include the further albedo feedback due to terrestrial snowline retreat, and the potentially very serious additional warming that may be created by unlocking methane from newly ice-free Arctic continental shelves.[27] The snowline and methane feedbacks cannot occur in the Antarctic – because of the lack of shallow shelves and the inflexible area of terrestrial snow cover. The Arctic amplification and greater Arctic feedbacks mean that, whatever the interactions between Antarctic sea ice and temperate oceans, it will always be the case over the next few decades that the Arctic will be determining the rate of global warming more than the Antarctic. In this sense the Arctic is a driver and the Antarctic can be thought of as a passive trailer in the global warming road race to oblivion.

13

The State of the Planet

So far we have focused on change in the polar regions, but now it is time to look at the planet as a whole and consider what state we are in.

First, there is no let-up in the rate of growth of greenhouse gas concentrations. Despite all the fine words of politicians and the efforts made by some countries to reduce their dependence on fossil fuels, the overwhelming effect of fuel-hungry economic growth in China and India is to continue to drive carbon dioxide concentrations ever upward. Given that levels, which have now reached 404 ppm (early 2016), are already too high for non-disruptive climate change, the fact that they are continuing to accelerate upwards with no let-up at all is profoundly distressing. They are not even beginning to slow. And let us remember that *all* of the CO_2 has a potential radiative forcing associated with it. Whether it is absorbed for a while in the ocean or in plants, it has by now been taken out of the ground, put into the climate system and is able to exert that radiative forcing, now or in the future, to heat the Earth. As we saw in Chapter 9, methane is an even more worrying gas. When its level in the atmosphere flattened off in the late 1990s people were relieved, and thought that some law of nature was asserting itself. It wasn't, and as of 2008 growth began again and is now approaching the growth rates of the 1980s. It is possibly significant that the resumption of methane growth coincided with large summer retreats of sea ice and associated warming of the Arctic shelf seabed; the link between Arctic offshore processes and global methane levels is becoming more and more firmly established, which means that there is worse to come.

Secondly, every planetary indicator looks negative. The human population, now 7 billion, is projected to reach 9.7 billion by 2050,[1] and 11.2 billion by 2100.[2] It is difficult to see how these numbers can be fed, given that we are experiencing large-scale climate disruption already, which is affecting the bread baskets of the world. Climate warming is reducing the area of cultivable land in places like sub-Saharan Africa, while theoretically improved yields at high latitudes cannot be realized because of extreme weather events. We are destroying forests. We are running out of water resources. And agriculture, which has to be an intensive, energy-hungry industry in order to feed so many people, is sensitive to shortages of vital raw industrial materials. The Nobel laureate Paul Crutzen, for instance, has drawn attention to the growing shortage of phosphorus, a vital element in the production of artificial fertilizers. The UN population predictions for 2100 are particularly worrying because they are split into continents: most continents show a large growth but one which can perhaps be coped with, while Europe shows a decline. However Africa shows a quadrupling in numbers, from 1.1 to 4.4 billion. Here are the figures:

Table 13.1. Current and projected populations by continent (millions of people)

	2015	2100
North America	358	500
South America	634	721
Europe	738	646
Asia	4,393	4,889
Oceania	39	71
Africa	1,186	4,387

Source: UN (2015), *World Population Prospects, the 2015 Revision*. Population Division, Dept. Economics and Social Affairs, United Nations

Since Africa cannot feed itself now, how will it cope with four times as many mouths, especially with global warming disrupting food supplies and causing desertification? The answer is that it won't. The rest of the world will have to feed Africa. Given that the rest of the world is likely to be obsessed with its own problems, one can foresee a shortage of compassion and a shortage of aid; the result will inevitably be famine on a massive scale. How will the world react to this evidence of its own selfishness? I quail at the thought of how nasty humanity may become, and of what excuses it will offer for inaction.

The population problem is not just one of food. Every human being is a carbon emitter, and so the problem of reducing total carbon emissions is made much more difficult if there are more people. Every human being needs space for someone to grow the food that he or she requires, so we see massive destruction of forests worldwide at a time when we desperately need more afforestation to reduce carbon dioxide levels. Every human being needs water to drink, and fresh water resources are getting scarcer, so that we may have to depend more on desalination, itself an energy-intensive process that releases carbon. It is difficult to deny the equation: more people = more carbon emission. Yet we seem to have forgotten the emphasis on the population explosion which concerned analysts of the global system in the 1970s, like the authors of *The Limits to Growth* (1972).[3] The problem hasn't gone away and it hasn't been solved, except for a while – by drastic means – in China.

Economically the world's rickety financial structure still requires perpetual growth in order to retain stability, with a banking system which is more and more obviously parasitic upon society. Within the present capitalist system, as practised by everyone including China, there is no way that a sustainable equilibrium society can be tolerated. Everyone knows that exponential growth in everything cannot continue and will lead only to disaster, yet every finance minister seeks to encourage economic growth to get his country out of the financial difficulties which he or his predecessors have created, with no thought of guiding this growth into sustainable channels.

Saddest of all is the personal paralysis that one sees in society. In the 1960s the young in the West were united in great crusades – against racism, against the Vietnam War – which showed that they really cared about the state of the world. Now, when the stakes are even higher and the need more urgent, they are passive. Voters of all ages, corporations and government bodies show a lack of concern with building a sustainable planet and a concern only with personal wealth and comfort. So long as we can consume luxuries, drive our cars and fly to holiday beaches for a few more years, we are quite willing to close our eyes to the certainty of future disease, poverty, warfare and crime, and ultimate food and resource depletion and starvation, all connected to the pressure of the rapidly changing climate system. The young are not listening or being inspired to action, and the old are not leading or teaching.

If we consulted Mr Micawber he would tell us that something is bound to turn up to save us from ourselves. But what might that be? Here are a few more or less unlikely scenarios:

- God might decide that it is time for the Second Coming (this is seriously advanced by a section of the US population as a reason to do nothing about climate change).
- UFOs might be real, and their sustained interest in us since 1947 might suggest that they plan to take over the planet for our own good.
- A miraculous device is invented which offers unlimited clean energy. This could be based on novel physics, like cold fusion, or on accepted physics, like a workable hot fusion system which has always been twenty years away.

Conversely

- We might be hit by a gigantic asteroid which wipes out all life.
- Some remote African forest region could breed a new strain of virus which sweeps the world and wipes out all, or most, human life.
- We could drift into a major nuclear war.

In my view a policy of waiting for something to turn up is far more likely to produce something bad than something good. Our salvation therefore lies in our own hands and our own actions.

WHAT CAN WE DO?

Emission reduction

In the past, and even today, green organizations have emphasized what we can do as individuals to mitigate climate change by reducing our carbon emissions. We can recycle our rubbish, insulate our homes, drive smaller cars, eat more vegetables and less meat. All these help, and also instil a sense of global civic virtue, of being aware of the needs of the global village as opposed to our own individual desires. But if every person in the UK applied every possible energy-saving measure to their normal lives, the result (from those that have tried it) is a reduction of only about 20 per cent in energy use. Useful, but, as the late Professor Sir David MacKay, the UK Government's chief scientific adviser on energy and climate, said, 'If everyone does a little, we will achieve only a little.'[4]

There is no doubt that to achieve more than a little, political decisions have to be made on energy production, which means that political courage must be shown by governments. Here despair sets in when one considers the history of the UNFCCC (United Nations Framework Convention on Climate Change) discussions, where the early optimism of the Kyoto Protocol (1997) gave way to the terrible failure of the Copenhagen (2009) and Durban (2011) meetings. Sadly, a typical politician's first response to the climate change crisis is to only quote predictions for this century, or even less, and to assume that climate change stops as soon as the IPCC graphs go beyond the 2100 limit. Britain's own past Secretary of State for the Environment, Food and Rural Affairs, Owen Paterson, said on 29 September 2013, with astounding complacency,

> I think the relief of this latest report is that it shows a really quite mod-
> est increase, half of which has already happened. They are talking one
> to two and a half degrees.[5]

Firstly, of course, 'they' were not the IPCC itself but an ignorant newspaper report on which he apparently relied for his knowledge. The 1–2.5°C is actually forecast for 2050. The 'half of which has already happened' demonstrates that he imagined that climate change stops at the end of the IPCC projections instead of going on. And of course the word 'relief' is the real giveaway; it was also, no doubt, a relief that he thought that he could get away without taking any action at all.

A typical politician's second response is that we can reduce our carbon emissions some time in the future (typically '30 per cent by

2032' or some suchlike figure) and thus stop climate change from getting out of control. This neatly lets current politicians off the hook. But it is untrue. For a start, the CO_2 already put into the atmosphere has a flywheel effect – a molecule of CO_2 lasts for much more than 100 years in the climate system and the world has yet to catch up on the potential warming of existing CO_2 (maybe only half has been 'realized'). So reducing our emissions in the future is much less useful than reducing our emissions now, and reducing our emissions now is less useful than actually reducing carbon levels. The most useful things to do would be actually to reduce the amount of CO_2 in the atmosphere, by carbon capture and storage or other technology yet to be invented; to completely cease to emit CO_2, for example by switching 100 per cent to nuclear power, which public opinion makes impossible; or to use technology to mask warming, i.e. put a sticking plaster on it, by geoengineering, buying us a little time. Nothing else can save us from serious consequences, although of course CO_2 reduction is still absolutely necessary. In this case the so-called 'green' organizations, such as Greenpeace and WWF, are unhelpful to humanity because of their opposition to both nuclear power and geoengineering.

The ratchet effect of carbon is rather like the ratchet effect of human population. Put very crudely, the 'natural' level of CO_2 in the atmosphere during interglacial periods is 280 parts per million (ppm) and thus of the present level of 404 ppm, more than 120 ppm has been put there by Man burning fossil fuels. Supposing we stopped emitting CO_2 altogether, all of a sudden. How fast would CO_2 levels go down? Well, with the survival time of added CO_2 in the Earth's energy system being at least 100 years, we might expect a maximum of 1 per cent of the added CO_2 to 'fall out' of the system per year, so CO_2 levels would diminish only 1.2 ppm in the first year of carbon abstinence. It will take forty-five years to bring the level down to the 350 ppm which most scientists think is 'safe'. Similarly, with the human population being 7 billion and an average lifetime of, say, seventy years, if humans completely ceased to reproduce it would take ten years for the population to diminish to 6 billion by natural decline. So if a food production crisis hits due to climate change and reduces our capacity to feed people by a billion, it will be impossible to match these new lower levels of food production quickly by birth control alone – nature will instead inflict mass starvation on us.

If we continue on our present path, eventually all hydrocarbons in the Earth will be extracted and burned, so our love-orgy with oil will have to come to an end. But by that time global warming will have become so extreme that life will be insufferable, if not impossible. We need a new Manhattan Project to clean up our atmosphere, an effort by the world greater than any effort that it has ever made, and it must be worldwide because we all breathe the same air. In the absence of such an effort the effects of climate change will become very apparent quite a short time into the future – in twenty or thirty years the world will be a different and much nastier place than it is now. There will never be another era for Man like the one that ended with the economic crisis of 2007. People will need to consider their personal futures and will try to live in cool countries like Norway or Canada, with low populations and many resources. This leads to the serious question, if it *is* now too late for us to preserve our planet by reducing or eliminating carbon emissions, because we have left it too late to start the process and because we live in a society in which high carbon emissions are 'built in' to the social and physical fabric, what can we do? There are only two possibilities: use technical methods to reduce the rate of warming while allowing CO_2 levels to continue to increase; or develop even more advanced technical means to actually take CO_2 out of the atmosphere.

The Royal Society, in its Geoengineering Report of 2009,[6] defined these two approaches:

- *Solar radiation management (SRM)* attempts to offset the effects of increased greenhouse gas concentration by causing the Earth to absorb less solar radiation.
- *Carbon dioxide removal (CDR)* addresses the root cause of climate change by removing greenhouse gases from the atmosphere.

Let us start with solar radiation management, the 'sticking plaster' approach of finding ways to reduce the rate of warming even as we continue to emit CO_2. These methods are classed generically as *geoengineering*.

Geoengineering

Buying precious time while we seek permanent solutions to the climate crisis requires the skills of technologists, whose abilities have never been more urgently needed in the service of mankind. Geoengineering comprises a suite of techniques to artificially lower surface air temperatures, either by blocking the sun's rays directly or by increasing the albedo of the planet so as to change the radiation balance. For the Arctic, the aim of both SRM and CDR must be to bring back the ice that we've lost, and in this way halt the loss of offshore permafrost and reduce the likelihood of a giant methane release. To achieve this, we need to not merely *slow* the pace of warming, but to *reverse* it. Let us look at the different ideas proposed, how effective they are likely to be, and the political difficulties.

SRM is the rapid 'sticking plaster' which can be implemented quickly at moderate cost. It does not deal with CO_2 levels, and so phenomena such as acidification of the oceans, which depend mainly on CO_2 levels rather than temperature, will continue apace, with serious consequences for bleaching of coral reefs, for shellfish survival, and in fact for the entire marine ecosystem. So SRM does not let us off the hook about reducing CO_2 levels as well.

Two major types of SRM methods have been proposed to date. In 1990 John Latham at the University of Manchester proposed 'whitening' low-level clouds by injecting very fine sprays of water particles into them.[7] This increases the cloud albedo and causes them to reflect more incoming solar radiation. The brilliant marine engineer Stephen Salter at Edinburgh University designed the systems to carry out this injection.[8] Others have proposed the injection of solid particles at high elevations from balloons or jet aircraft afterburners, which would form aerosols that reflect incoming radiation.

Marine cloud brightening (MCB) involves increasing the amount of sunlight reflected back into space from the tops of thin, low-level clouds (marine stratocumuli, which cover about a quarter of the world's oceanic surface), thereby producing a cooling effect. If we could increase the reflectivity by about 3 per cent, it is estimated that the cooling will balance the global warming caused by increased CO_2

in the atmosphere. To do this we need to spray sea-water droplets continuously into the cloud. Salter developed plans for a novel form of spray-droplet production, and designed an unmanned wind-powered vessel that can be remotely guided to regions where cloud seeding is most favourable (Plate 30). Instead of sails, such a vessel could use a much more efficient motive power technique – Flettner rotors. These spinning vertical cylinders mounted on the deck are named after their inventor, the German Anton Flettner, and make use of the Magnus effect, whereby a spinning vertical cylinder has a pressure difference across its sides which gives a force at right angle to the wind direction. Flettner rotors were used for ships in the 1920s and have been revived today as a way of reducing fuel consumption at sea. The rotorship houses the spraying system which sprays sea-water droplets from the top of the rotors into the cloud base. The power required for spraying and communications comes from electricity generated by current turbines built into the vessels. The key to the design is the fine nozzle which produces particles of the required diameter of about a micrometre (a millionth of a metre), such that when the droplet evaporates in the atmosphere it produces a tiny salt particle which is of just the right diameter (a nanometre or so) to brighten the cloud. This makes use of the so-called Twomey effect, that a mass of tiny particles in a cloud is brighter than the same mass of larger particles. This effect has been observed from ships leaving the equivalent of a con-trail of brighter clouds, visible from space (Plate 31). Several hundred vessels distributed worldwide would be needed to achieve the aim, but the total cost, while substantial, is small compared to the massive costs of global warming to the planet – billions of dollars per year compared to global warming costs of trillions. A huge advantage of the plan is that it is ecologically benign, the only raw material required being sea water. The amount of cooling could be controlled, via satellite measurements and a computer model, and if an emergency arose, the system could be switched off, with conditions returning to normal within a few days.

Much work is needed before a cloud brightening system could be operational. We would have to complete the development of the technology, and conduct a limited-area field experiment in which the reflectivity of seeded clouds is compared with that of adjacent unseeded ones. We would also have to perform detailed analyses to establish whether there might be serious or harmful meteorological or climatological ramifications (such as reducing rainfall in regions where water is scarce) and, if so, to find a solution for them. One question is, must we spray worldwide and thus achieve global effects or, if we specifically desire a regional outcome, can we spray in particular locations or at particular times of year? The very urgent question of cooling the Arctic comes to mind. If it is the open water over the Arctic shelves in summer which allows warming of the sub-sea permafrost and a potential methane catastrophe, can we ward this off by bringing back the summer sea ice without necessarily having to cool the entire planet?

This regional question was tackled by John Latham and colleagues in 2014.[9] We found that indeed it is possible to focus the cooling on the Arctic and cause some advance in the sea ice limits, especially in the Beaufort and Chukchi seas, although there may be compensating problems such as reduction of rainfall in sub-Saharan Africa. Clearly this is cautiously promising. In an earlier study, it was estimated that spraying over 70 per cent of the marine global cloud cover could eliminate the warming due to CO_2 doubling and halt losses of sea ice.[10] The cloud brightening is basically reducing the radiation falling on the ocean surface, so, if regionally targeted, this might also have beneficial effects in reducing the vigour of hurricanes (which depend on sea surface temperature) and the rate of bleaching of coral reefs (which depend on water temperature as well as ocean acidity). Finally the Antarctic sea ice could also be affected: the 2014 study showed that global seeding will increase the Antarctic sea ice area and also cool at source the subsurface currents which at present are threatening to cause the Thwaites and Pine Island glaciers to collapse, which could cause a serious, 3-metre rise in global sea level if it occurred suddenly.[11] So MCB may offer not only global relief with respect to warming, but also relief from regional threats, especially in the polar regions.

Stephen Salter mapped out a development plan which estimates that taking a cloud brightening system to full operational effectiveness would cost £73 million in research and development costs, a fortune in terms of normal science budgets but a pittance in terms of the urgent global need. If Britain were serious about fighting global change, this would be an area where it could take a lead.

Aerosol injection is the second large-scale geoengineering method that has been proposed.[12] Some of the implications were studied in a recent project supported by the UK Government called SPICE (Stratospheric Particle Injection for Climate Engineering), although support was withdrawn before the scientists could actually try a system out. The idea is to disperse a large mass of aerosols, tiny particles, into the stratosphere at high level, so that they can directly reflect sunlight back into space. The injection would have to be continuous, as the aerosol gradually falls out of the upper atmosphere.

The original ideas called for the creation of a stratospheric sulphate aerosol cloud, either through the release of a so-called precursor gas – sulphur dioxide (SO_2) – or through the direct release of sulphuric acid (H_2SO_4). If SO_2 gas is released it will oxidize in the upper atmosphere and dissolve in water to form sulphuric acid in droplets far from the injection site. This does not allow control over the size of the particles that are formed but the gas is fairly easy to release. If sulphuric acid is released directly then the aerosol particles would form very quickly and, in principle, the particle size could be controlled to optimize the climatic effect. If the aerosol is injected into the lower stratosphere it will remain aloft only for a few weeks or months, as air in this region is predominantly descending, so to ensure a longer lifetime of years, higher-altitude delivery is needed.

How might this be done? Delivery systems suggested include artillery shells, high-altitude aircraft, or high-altitude balloons, either supporting vertical pipes from the ground or rising freely until they burst, with the precursor gas inside them. The cheapest systems appear to be existing tanker aircraft such as the US KC-135 or KC-10 military tankers, only nine of the larger KC-10 aircraft being required to deliver 1 Teragram (1 million tons) of sulphur dioxide per year at three flights per day. Sixteen-inch artillery shells are comparable in cost, as are a huge number of small balloons in which hydrogen sulphide (H_2S), another possible precursor gas, is mixed with hydrogen to give a buoyant balloon which bursts when it reaches the stratosphere; 37,000 commercial balloons would be needed per year. The systems are simpler than for marine cloud brightening, but the quantities involved are very large and the chemicals have to be hoisted high into the atmosphere.

We know that high-altitude particles can indeed affect climate, because of volcanic eruptions – the eruption of Mount Pinatubo in 1991, for example, produced a noticeable global cooling for three years afterwards. The cost would be within bounds, being estimated as 25–50 billion dollars per year to fully counteract Man's additions of carbon dioxide, according to Paul Crutzen who was an early proponent of this idea.[13] However, many potential problems have been identified. Rainfall would be reduced, which might have a serious impact on the Asian and African monsoons; there might be an increase in the rate of ozone destruction, leading to a regrowth of the ozone hole; it is difficult to predict how the cooling will be distributed worldwide, so some countries may experience less cooling than others, or even warming, and so on. Behind it all lies an unease with the idea of injecting large quantities of an undeniably toxic chemical into the upper atmosphere; marine cloud brightening with sea water particles sounds positively benign by comparison. One relentless opponent of aerosol injection, Alan Robock of Rutgers University, has recently changed his views, however; he co-authored a 2016 paper[14] which showed that the aerosol cloud would not only reduce direct radiation reaching the ground, it would enhance diffuse radiation, which would combine with the cooling to produce an increase in plant photosynthesis rates. This increase in plant growth would itself play a role in reducing carbon dioxide levels in the atmosphere, an unexpected extra benefit.

A few other geoengineering techniques have been proposed. One is the space reflector, a very large mirror or system of mirrors in orbit, to reflect large amounts of sunlight into space. However, nobody has come up with a feasible plan that could assemble anything like this in orbit at anything other than colossal cost.

Carbon drawdown

I have explained why carbon emission reduction is unlikely to happen, at least not fast enough, and if it is done too slowly, as it will be, it will leave a legacy of excess CO_2 in the atmosphere which will continue to drive future warming. Geoengineering can counteract the impact of carbon dioxide and methane on the atmosphere, but at the cost of leaving the CO_2 to continue its work in acidifying the ocean, which could ultimately destroy our marine ecosystem (and thus our global ecosystem, since the ocean makes up 72 per cent of the planet's surface area). My bleak conclusion is that in the end (and that end may be close) we have to find a way of taking CO_2 out of our planetary system if we are to defeat global warming and save our civilization. How do we do it?

First of all, we must take the whole problem seriously. In fact, it is the most serious question facing the world – can we bring ourselves back from the brink of runaway climate change and retain the basis of viable life for the human race? Or must we grapple hopelessly with accelerating climate change which makes large parts of the planet unlivable? In this respect one of the most shameful failures has been that of the IPCC. In its Fifth Assessment report in 2013, the IPCC recognized that the only way to a viable climate is to follow the RCP2.6 route. I have already expressed my suspicion that the 'RCP' formulation of radiative forcing conceals the realities of what is needed to avoid disastrous climate change (see Chapter 7). But the result is a paradox, unstressed by the IPCC: the only way to save ourselves is to follow the RCP2.6 route, and the only way to do that is to actually take CO_2 out of the atmosphere, since we will very shortly reach the CO_2 concentration (421 ppm) which is the ceiling for 'acceptable' climate warming. We are certain to pass that limit without even noticing it in about a decade, so fast are CO_2 levels rising, and beyond that point our only hope is actual carbon removal. The IPCC knows this but ignores the problem of how we might actually remove CO_2. A further crucial component of CO_2 removal is the impact that large-scale application of it could have on ecosystems and biodiversity. This has to be studied on an international scale before we begin attempts at large-scale CO_2 removal; again, this is ignored by the IPCC.

Two possible techniques have assumed prominence recently.[15] They are *bioenergy with carbon capture and storage* (BECCS), and *afforestation*. BECCS involves growing bioenergy crops, from grasses to trees; burning them in power stations; stripping the CO_2 from the resulting waste gases; and compressing the gas into a liquid for underground storage. Afforestation – planting trees – also relies on photosynthesis to remove CO_2 from the atmosphere; storage is achieved naturally, in timber and soil. If we are to limit the global temperature rise to 2°C, we need to remove some 600 gigatons of CO_2 by the end of this century. Using BECCS, this would require crops to be planted solely for the purpose of CO_2 removal on between 430 million and 580 million hectares of land – around one-third of the current total arable land on the planet, or about half the land area of the United States. This is clearly impossible, unless we can achieve remarkable increases in agricultural productivity which greatly exceed the needs of a rapidly growing global population. It is more likely that we will need that arable land to feed people (and in any case it will probably be less productive because of the extreme weather effects arising from Arctic change). BECCS would have to use primary forest and natural grassland, which equally we cannot spare because afforestation is itself one of the possible ways of removing CO_2. These wild places also contain the last strongholds of vast numbers of threatened terrestrial species, the loss of which might be disastrous for the continued survival of the planetary ecosystem. A further fundamental concern is whether BECCS would be as effective as is assumed at stripping CO_2 from the atmosphere. Planting crops at such a scale could involve more release than uptake of greenhouse gases, at least initially, as a result of land clearance, soil disturbance and increased use of fertilizer. When such effects are taken into account, the maximum amount of CO_2 that can be removed by BECCS (under the RCP2.6 scenario) has been estimated to be 391 gigatons by 2100, about 34 per cent less than the amount assumed to be needed to keep the temperature rise below 2°C. If less optimistic assumptions are made about where the land for bioenergy crops can come from, the net capture by 2100 comes down to 135 gigatons. So it already looks as if BECCS cannot do the job alone. On top of all this, we would be planting bioenergy crops into a world of changing climate: what will their water requirements be in a warmer world? How will they compete with food production if overpopulation really leads to a race for cropland? And (as with other techniques) how do we capture and where do we store the carbon dioxide?

Afforestation sounds a gentler way of taking CO_2 out of the atmosphere, since we don't have to put it anywhere. Everyone assumes that an increased forest cover is environmentally desirable – even while we are busy chopping down the Amazon and Southeast Asian forests for their hardwoods and to grow soya beans or keep cattle. How do we grow more forest when all the pressure is towards forest loss? Afforestation can also involve the loss of natural ecosystems, if we are replacing natural forest by managed monospecies forests. We are only at the beginning of the proper study of key forest species whose loss could be disastrous either for preserving our global ecosystem, or for keeping down serious pests like the bark beetle.[16] A third of new drugs are developed from forest plants. And planting swathes of managed forest will cause complex changes in cloud cover, albedo and the soil–water balance, through changes to evaporation and plant transpiration. One undesirable change is occurring with the northern boreal forest. With global warming the treeline is moving north, which one might think of as a good thing except that during the season of snow cover a terrain covered with tree foliage (bare branches or evergreen needles) is darker than flat snow-covered grassland or tundra, so the overall albedo is reduced and again there is a net warming effect. Systematic use of afforestation will involve cutting down the trees (and storing the wood) when they have reached a certain stage of growth, followed by replanting; this will not work if increased fires, droughts, pests and disease cause the trees to die and fall before harvesting.

Many other ideas have been proposed for CO_2 removal by biological, geochemical and chemical means. For all such schemes, modelling theoretical potential can give a completely different picture from that obtained when environmental impacts – not to mention practicalities, governance and acceptability – are considered. A case in point is the long history of discussion, research and policymaking on *ocean fertilization*, another CO_2-removal technique. When a link was first made between natural changes in the input of dust to the ocean, ocean productivity and climatic conditions, there were high expectations of how effective ocean fertilization might be as a way to avoid human-driven global warming. During the 1990s, researchers postulated that for every ton of iron powder added to sea water, tens of thousands of tons of carbon (and hence CO_2) could be fixed by the resulting blooms of phytoplankton. This estimate has been whittled down over the years and after fourteen small-scale field experiments, with the realization that most of the CO_2 absorbed by such blooms – stimulated either by adding iron or other nutrients to sea water, or by enhancing upwelling through mechanical means – is released back into the atmosphere when the phytoplankton decompose. Moreover, a large-scale increase in plankton productivity in one region (across the Southern Ocean, say) could reduce the yields of fisheries elsewhere by depleting other nutrients, or increase the likelihood of mid-water deoxygenation. Such risks have resulted in the near-universal rejection of ocean fertilization as a climate intervention, through bodies such as the Convention on Biological Diversity (CBD).

More recently, other ocean-based CO_2-removal techniques have been suggested, such as the cultivation of seaweed to cover up to 9 per cent of the global ocean. The specific environmental implications of this method have yet to be assessed. Yet such an approach would clearly affect, and potentially displace, existing marine ecosystems that have high economic value, especially in shallow waters.

Back on land, other techniques include those to increase the amount of carbon sequestered in the soil, for example by ploughing in organic material such as straw, reducing ploughing (to limit soil disturbance) or adding *biochar*. Biochar has an interesting history all of its own, because of the efforts of a band of enthusiastic supporters to persuade the world that this is the answer to global warming. Crop materials or farm wastes are digested by a process called pyrolysis, which produces a liquid and leaves a charcoal-like spongy material that can be dug into soil and allegedly gives it special properties. It is never properly explained how all this disposes of CO_2. Another idea among some enthusiasts is to enhance weathering, which involves the absorption of CO_2 from the atmosphere by certain *silicate rocks*, especially olivine. The material has to be crushed to provide as much surface area as possible, and thus would need to be spread on beaches and other surfaces as a fine white sand. A slow chemical reaction then ensues which absorbs CO_2 and emits oxygen. It is true, as the enthusiasts say, that this is the chemical process in the early Earth which first released oxygen from rocks. Yet to reduce the amount of CO_2 in the atmosphere by 50 parts per million, to get us back to 350 ppm from our current 400 ppm, 1–5 kilograms per square metre of silicate rock would need to be applied each year to 2–6.9 billion hectares of land (15–45 per cent of Earth's land surface area), mostly in the tropics. The volume of rock mined and processed would exceed the amount of coal currently produced worldwide, with the total costs of implementation estimated to be between \$60 trillion and \$600 trillion, far more than geoengineering techniques. Like geoengineering, the application would have to be continuous, since once the chemical reaction is over the rock is of no more use and must be covered with fresh layers. Clearly the whole thing is unfeasible. Yet it is crucial to know more about the permanence of carbon storage for biologically based methods, and the environmental impacts that might result if such approaches are used at vast scale, so a wide range of research is needed.

All these methods therefore have serious, if not fatal, drawbacks. We are left with something that has yet to be invented, but which ought to be the subject of a research programme on the scale of the Manhattan Project, *direct air capture* (DAC). DAC means pumping air through a system which removes the CO_2 and either liquefies it and stores it or turns it chemically into something else, hopefully something useful. When I say it has 'yet to be invented', I mean that a system which is not impossibly expensive has yet to be invented. DAC can, in principle, be undertaken by passing air through anion-exchange resins that contain hydroxide or carbonate groups, which, when dry, absorb CO_2 and release it when moist. The extracted CO_2 can then be compressed, stored in liquid form and deposited underground using carbon capture and storage technologies. The huge operational costs for DAC cover a similar range to those estimated for enhanced weathering, at the moment amounting to more than \$100 per ton of carbon, although a recent (2016) breakthrough promises \$40 per ton. The extraction process would also need land and probably water, and, as for BECCS, there is a risk of CO_2 leaking out of geological reservoirs. Such risks can be minimized by storing the liquid CO_2 beneath the sea or by using geochemical transformation, which involves *in situ* reactions between CO_2 and certain rock types. In theory, cooling (rather than chemistry) to liquefy the

CO_2 could also be used to remove CO_2 from ambient air. The technical feasibility, costs and potential environmental impacts of this approach – which could involve setting up plants on high polar plateaus such as Antarctica or Greenland – have yet to be investigated. Since my own belief, based on the above reasoning, is that Direct Air Capture is all that we are left with as a way of maintaining the world in anything like its present condition in the long run, then if we carry out serious research on the scale of the wartime Manhattan Project we may be able to bring down the cost in the same way as solar photovoltaic energy has plummeted in cost in recent years.

A valid criticism of geoengineering or carbon removal is that it encourages us to do little or nothing about reducing our carbon dioxide levels, and that our urgent actions should focus on emissions reduction and not on an unproven '*emit now, remove later*' strategy. But the unfortunate reality is that the global population, especially in the West, is extraordinarily reluctant to give up the comforts and conveniences of living in a fossil fuel world. We will eventually, because we will have to. But we don't see why we should just now. Just one more Ryanair flight, and isn't that SUV a good way of getting the children to school? But even if a drastic and immediate effort is made to cut emissions, significant geoengineering and CO_2 removal operations will need to begin around 2020, with up to 20 gigatons of CO_2 extracted each year by 2100 to keep the global temperature increase below 2°C. We need to know if that is feasible, in order to answer the next question.

CAN THE 2015 PARIS AGREEMENT SAVE US?

In December 2015, in Paris, the 195 parties to the United Nations Framework Convention on Climate Change (UNFCCC) achieved a historic agreement at the COP21 (twenty-first Conference of the Parties) meeting. They agreed to achieve a stabilization of greenhouse-gas concentrations some time between 2050 and 2100. This commitment (signed by the various nations in April 2016) is intended to limit the increase in global average temperature above pre-industrial levels to 'well below 2°C' – and preferably to 1.5°C. A balanced greenhouse-gas budget either requires that industry and agriculture produce zero emissions or necessitates the active removal of greenhouse gases from the atmosphere (in addition to deep and rapid emissions cuts). In most modelled scenarios that limit warming to 2°C, several gigatons of carbon dioxide have to be extracted and safely stored each year. For more ambitious targets, tens of gigatons per year must be removed. Hence the link with our discussions in this chapter.

The provisions of the agreement can be summarized as follows. The governments agreed:

- a long-term goal of keeping the increase in global average temperature to *well below* 2°C above pre-industrial levels;
- to aim to limit the increase to *1.5°C*, since this would significantly reduce risks and the impacts of climate change;
- on the need for *global emissions to peak as soon as possible*, recognizing that this will take longer for developing countries; and
- to undertake *rapid reductions thereafter* in accordance with the best available science.

Before and during the Paris conference, countries submitted 'intended nationally determined contributions' (INDCs). These are individual national commitments to reduce carbon emissions. These are not yet enough to keep global warming below 2°C, but the agreement traces a way to achieve this target, since the governments agreed to:

- come together every five years to *set more ambitious targets* as required by science;
- *report* to each other and the public on how well they are doing to implement their targets;
- track progress towards the long-term goal through a robust *transparency and accountability* system;
- strengthen societies' ability to *deal with the impacts* of climate change; and
- provide continued and enhanced international *support* for adaptation *to developing countries*, with a goal of $100 billion per year up to 2025.

The agreement also:

- recognizes the importance of averting, minimizing and addressing *loss and damage* associated with the adverse effects of climate change; and
- acknowledges the need to *co-operate* and enhance the *understanding, action and support* in different areas such as early warning systems, emergency preparedness and risk insurance.

What does all this mean? The positive aspects are very clear. It is the first truly global climate agreement to be achieved. It has brought the US back into the process, and engaged India, China and other large emitters. It has changed the 'storyline', in that, instead of the acrimony and wriggling that occurred at previous meetings in Copenhagen and Durban, where countries tried to do the absolute minimum or nothing at all, everyone is now keen and honest and devoted to a single critical goal. There is a chance for genuine international partnership and not just interaction. So in many ways the agreement was a diplomatic and political triumph, which is unequivocally positive in the light of what has gone before.

But can it save us? Let's look at some of the things that are not in the agreement. First of all, it is inconsistent with a pathway to a safe climate. The aim is to keep warming below 2°C, but the INDCs presented so far, even if fully honoured, will leave us with a warming of at least 2.7°C. There is no possibility of getting near 1.5°C except with massive use of geoengineering and carbon removal, not mentioned in the agreement, which deals only with emissions. There is no mention either of aviation, which is a major factor in global warming. There are no plans for immediate actions, and no date set for achieving carbon balance except 'between 2050 and 2100' which is dangerously vague, as the later date would imply carbon balance being achieved at a high carbon dioxide level. In short, the agreement is very dependent on national goodwill and honesty to make it work, though the review meetings should help. Fundamentally, climate change is a 'stock-flow' problem: the rise in temperature is closely associated with the accumulation of emissions over time (the stock), but we are only able to control the rate at which emissions are emitted or removed from this point forward (the flow). Our planet has already accumulated a large amount of emissions, and to stabilize or reduce the atmospheric concentration of greenhouse gases, current emissions must be reduced by at least 90 per cent, which requires the implementation of emissions reduction technology.

The agreement is thus a huge step forward, but only a step. It gives us an agreed target, but does not show anyone how to achieve that target. I believe that the stabilization target can in fact be achieved only by interventions in the form of geoengineering and carbon draw-down technology, and that if the world strives to restrain global warming to 1.5–2°C by carbon emission cuts alone, the result will be an embarrassing failure. Now is the time to focus on bringing in these new technologies, before emission cut failures lead to squabbling and a breakdown of the agreement. The Paris agreement is a step that should have been taken ten or twenty years ago, and we should by now have moved on to the serious business of really tackling climate change.

14

A Call to Arms

The discovery in 2015 of the very high long-term climate sensitivity of the planet to greenhouse gases[1] is of the utmost importance in clarifying what should be our priority as human beings in the crisis that faces us. It shows that the *existing* level of carbon dioxide in the atmosphere is sufficient to cause unacceptable amounts of warming in the future. We no longer have a 'carbon budget' that we can burn through before feeling worried that we have caused massive climate change. We have burned through the budget and are causing the change now.

Therefore, *it is not enough to reduce carbon emissions*. Twenty or thirty years ago, when global warming was first recognized as a severe threat, a serious, concerted effort by the international community to reduce fossil fuel use, and to switch to renewable energy sources, including nuclear, might have been enough to slow global warming to the point where the Earth could experience a soft landing at a temperature that is not dangerously high. But governments and peoples alike were too short-sighted, ignorant and greedy to make the necessary changes. Nothing could be hoped for when countries like China and India were accelerating their fossil fuel use, especially coal. By now it is too late. The CO_2 levels in the atmosphere are already so high that when their warming potential is realized in a few decades, the resulting temperature rise will be catastrophic.

To avoid such a fate, we must not only go to *zero emissions*, we must actually *remove carbon dioxide from the atmosphere*. Only in this way can we avoid dire consequences. But as I showed in the last chapter, this is extremely difficult. The techniques that have been proposed and developed so far are expensive, costing about $100 per ton of carbon, and we have to remove every year an amount *exceeding* our emissions, which are 35 billion tons. A vast and urgent research project is needed to develop cheaper methods; improved catalytic methods have been proposed which might bring us to $40 per ton. It ought to be possible to achieve a cheaper price, and it is also a psychologically more congenial approach than asking people to stop their carbon emissions immediately in a world whose infrastructure is set up to encourage fossil fuel use. In the USA especially, a great project to develop new technology for removal of airborne carbon is the kind of challenge that the can-do mentality of America appreciates.

While these methods are being developed and implemented, we will need geoengineering to put a sticking plaster on the planet. I am fully aware that geoengineering does nothing about the causes of global warming, does nothing to ameliorate CO_2 impacts such as ocean acidification, may have side effects and an unexpected geographical distribution of impact, and requires constant application. But, without it, temperature rise, and the associated further feedbacks, will be too great to allow our civilization to continue.

We have destroyed our planet's life support system by mindless development and misuse of technology. A mindful development of technology, first for geoengineering, then for carbon removal, is now necessary to save us. It is the most serious and important activity in which the human race can now be involved, and it must begin immediately.

IMPROVING THE SCIENCE

Let us return for a moment from the global scale to the Arctic scale and look at how we could be improving our science, and in particular bringing economics into the physics. The worldwide nature of the costs of Arctic warming shows unequivocally that all countries, not just those in the far north, should be concerned about changes occurring in these regions. In costing the effect of offshore Arctic methane emission we have examined one environmental impact out of many. More must be done to establish the economic consequences of other Arctic feedbacks and where they will hit hardest. The full financial impact of a changing Arctic is likely to be considerably greater than our initial estimate of that due to methane release, which was itself very high.

First, we need computer models that better integrate physical changes to the Arctic and economic impacts across space and time that are not yet explicitly addressed by the PAGE model. The models should incorporate feedbacks such as linking the extent of Arctic ice to increases in Arctic mean temperature, global sea level rise and ocean acidification. They should also include feedbacks that are not explicitly modelled in the current version of PAGE, such as the effects of black carbon deposits and of tundra permafrost melt. They should link the extent of Arctic ice to increases in Arctic mean temperature, and then link economic impacts, such as increased shipping or global sea level rise, to Arctic ice extent. Such integrated models of the economic costs of Arctic change should disaggregate global impact figures into countries and industry sectors. This could help to raise awareness of particular risks in specific nations, such as small island states, or coastal cities like New York. These feedback links are not included in the current analysis, but could be in future work.

Secondly such integrated analyses – and those who undertake them – need to join global economic discussions. For example, the World Economic Forum (WEF) launched its new Global Agenda Council on the Arctic in the autumn of 2012, citing the need for informal dialogue among world leaders and recognizing that the Arctic is increasing in strategic importance both in terms of potential economic value (from shipping and minerals extraction[2]) and ecological vulnerability. Yet, in a TV discussion about the 2014 Davos meeting, the words 'climate change' were mentioned only once, and the topic was not discussed at all by the pundits present.

Without denying the economic potential of the Arctic region, a rigorous economic analysis is clearly needed to recognize global impacts and costs from Arctic change. The WEF could help to kick-start investment in this new kind of integrated and systemic approach to economic assessment – one that considers how physical changes to ecosystems such as the Arctic will affect the global economy. The WEF could also use its significant convening power to ask world leaders to consider the full spectrum of costs and benefits from a changing Arctic, and to redirect economic attention from short-term economic gains from shipping and extraction to what appears to be an economic and ecological time-bomb. Already we have seen (Chapter 9) that a single feedback could carry the enormous price tag of $37–60 trillion over a century, with most of the impacts occurring in poorer countries, set against a world economy of approximately $70 trillion,[3] so the costs of Arctic change carry enormous risks to our global economic foundations. We can expedite change by including such costs in the WEF's Global Risk Report and the IMF's World Economic Outlook,[4] neither of which currently recognize these potential economic threats emerging from the Arctic.

Thus in establishing the scientific needs for climate change mitigation in the Arctic, we actually need to develop a new scientific approach, an *integrated Arctic science*. Integrated Arctic science is a strategic asset for human economies because what happens in the Arctic has critical effects in our biophysical, political *and* economic systems. Without this recognition, economists and world leaders will continue to miss the big picture.

DANGERS OF WAR

As I started to write this book in 2013, the world remembered the death of John F. Kennedy half a century earlier. This brought renewed attention to the Cuban Missile Crisis of 1962 and how near the world came to nuclear war. I can remember, as a fourteen-year-old boy, watching the BBC News on 27 October 1962 and suddenly realizing, as did my mother and father, that we might not wake up the next morning, that the safe little world of our semi-detached house in Essex could easily vanish and be turned into ash, along with ourselves and most of the British population. All because of a little island whose conduct America was exercised about. The wisdom and restraint of Kennedy and Khrushchev are praised today, as they were in 1962, but to deliberately bring the world to the brink of destruction, just over the status of Cuba, does not speak of wisdom – it speaks of madness. Today we congratulate ourselves that the Cold War is over and that this sort of confrontation will not recur, despite the vast stockpiles of doomsday weapons that the US and Russia continue to hold. But a host of new states now possess nuclear weapons, not just great powers with moderate policies but volatile nations such as Israel, North Korea and Pakistan, countries that possess nuclear weapons and seem very prepared to use them if their religious or political obsessions are challenged. The threat is greater than ever. A nuclear war would now probably start because of a bilateral issue, and climate change is bringing a host of new stresses which could create such an issue, from resource and water depletion to collapses in food production with the looming potential for starvation. Nuclear weapons have been invented, and the world will not get rid of them completely unless and until human nature changes, since there would have to be trust that the irrational nations would follow the rational ones in giving them up. Yet human nature has not changed, except perhaps for the worse. Aleksandr Solzhenitsyn described the twentieth century as the 'cave man century', and a new century that began with our illegal invasion of Iraq and a million pointless deaths can scarcely claim to be turning into a better one. But if we cannot get rid of nuclear weapons completely without changing human nature, and we cannot change human nature, then in the end nuclear weapons will be used. It could well be the global stresses produced by climate change that provide the spark for the flame that ends the human race, which is another crucial reason for tackling climate change, working together as a species rather than as a collection of mutually antagonistic nations. Time is running short to avoid major disruption to the planet, but it can be done. But if a nuclear war starts, time will have run out on the human race, instantly and completely.

THE BLACK TIDE OF DENIAL

When the phenomenon of global warming first became apparent to scientists during the early 1980s, there was a general optimism that, once the facts and the mechanisms were explained to the public and politicians, there would be overwhelming support for the necessary international actions to curb carbon emissions, switch to renewable energy, and save our planet from the worst excesses of global change.

Indeed, this process seemed to be under way. In the UK, the then prime minister, Margaret Thatcher, who had been trained as a chemist, showed an immediate understanding of the scientific principles involved and took up the need for international action on climate change as a principal task of the latter part of her premiership. She founded the Hadley Centre for Climate Research and Prediction at the UK Meteorological Office in 1990 and pressed for international action. Her recognition that the polar regions are of special importance was reflected in a message which she sent me in 1989, when I was in the Antarctic on an icebreaker, asking for a statement on polar change which she could give at the United Nations General Assembly. Here is what she said on 8 November 1989, ascribed to 'a British scientist on board a ship in the Antarctic Ocean':

> 'In the Polar Regions today, we are seeing what may be early signs of man-induced climatic change. Data coming in from Halley Bay and from instruments aboard the ship on which I am sailing show that we are entering a Spring Ozone depletion which is as deep as, if not deeper, than the depletion in the worst year to date. It completely reverses the recovery observed in 1988. The lowest recording aboard this ship is only 150 Dobson units for Ozone total content during September, compared with 300 for the same season in a normal year. That of course is a very severe depletion.'

> He also reports on a significant thinning of the sea ice, and he writes that, in the Antarctic, 'Our data confirm that the first-year ice, which forms the bulk of sea ice cover, is remarkably thin and so is probably unable to sustain significant atmospheric warming without melting. Sea ice separates the ocean from the atmosphere over an area of more than 30 million square kilometres. It reflects most of the solar radiation falling on it, helping to cool the earth's surface. If this area were reduced, the warming of earth would be accelerated due to the extra absorption of radiation by the ocean.

> 'The lesson of these Polar processes,' he goes on, 'is that an environmental or climatic change produced by man may take on a self-sustaining or "runaway" quality ... and may be irreversible.' That is from the scientists who are doing work on the ship that is presently considering these matters.

These are sobering indications of what may happen and they led my correspondent to put forward the interesting idea of a World Polar Watch, amongst other initiatives, which will observe the world's climate system and allow us to understand how it works.[5]

The United Nations established the Framework Convention on Climate Change (UNFCCC) by treaty at the Rio Earth Summit in June 1992. Earlier, in 1988, the World Meteorological Organization (WMO) and the United Nations Environment Programme (UNEP) had set up the Intergovernmental Panel on Climate Change (IPCC) which produced its first assessment in 1990.[6] In her inspiring speech, Mrs Thatcher recommended that the IPCC became a long-lasting organization producing further assessments. But then the political leadership flagged. In the UK, Mrs Thatcher was removed from office, for other reasons, in 1990, just when she was beginning to make an international impression on the problem. Her successors, Messrs Major, Blair, Brown and Cameron, had no scientific training, were often politically weak, and mouthed platitudes about leading the international effort on climatic change while actually doing little. In the US things were even worse, with the two presidents Bush actively opposed to any measure which might threaten the hegemony of the oil industry, of which they were beneficiaries, and even presidents Clinton and Obama, while they made inspirational speeches, actually did very little. The US could not bring itself to sign up to the Kyoto Protocol of 1997, which could have been a start in the process of reducing emissions, even though the international community bent over backwards to make it palatable to the US. In the very small print, for instance, the Protocol exempted military flying from emission controls; this was to please the US which does more military flying than the rest of the world put together, and involved everyone accepting the fiction that a carbon dioxide molecule emitted by a military aircraft has less effect on the planet than one emitted by a civil aircraft.

Worse than the continuing international inertia and the lack of political leadership, as evidenced by the failure of the UNFCCC 'summits' in Copenhagen and Durban, is the fact that an insidious opposition to taking action on climate change is now being fomented by well-financed groups of malevolent people and organizations. These organizations focus on planting stories in the media and persuading timid or ignorant politicians that we cannot afford to do anything about global warming, even if it actually exists. Their aims and methods are exactly the same as those of tobacco industry lobbyists – to sow doubt about the harmfulness of the impacts to the point that ordinary people become confused and are willing to tolerate inaction. They don't have to persuade people that climate change is not happening – just sow doubt, and since action to save the world involves effort, cost and discomfort, it is always tempting to latch on to a statement that we don't really need to do anything at all. A powerful book about this movement is called *Merchants of Doubt*.[7]

The denial movement, which is now estimated to be funded by elements of the oil industry and secretive industrialists to the tune of $1 billion per year, has acted in two ways. First, vicious personal attacks on the careers of climate scientists whom they target as being genuine experts on climate change and thus likely to be outspoken. They achieved their first major success in 2002, when, as a result of a scandalous memo by a certain Randy Randol of ExxonMobil to the White House, President Bush instructed the US delegation to the IPCC to remove Professor Robert Watson as Chair of the IPCC and to have him replaced by someone more pliable. The US could do this because it is the largest single funding source for the IPCC. Watson, an energetic and brilliant climate scientist, was seen as possessing dangerous fervour, especially when in the late 1990s he announced that the latest revision to climate models, involving an improved treatment of the carbon cycle, suggested that the global climate was warming a third faster than previously thought. He was replaced by Rajendra K. Pachauri, a mild-mannered Indian who, nevertheless, gradually become radicalized by the sheer magnitude of the threat facing the world, leading to a Nobel Peace Prize for the organization (shared with Al Gore) in 2007. As an IPCC author dating back to 1990 I received an impressive certificate for 'contributing to the award of the Nobel Peace Prize', signed by Pachauri and the IPCC secretary, 'R. Christ', the latter signature giving it almost a holy feel. I also received a plastic lapel badge, but its level of tackiness was such that neither I nor any scientist I know has ever worn it.

Next on the deniers' hit list was James Hansen, until recently Director of the NASA Goddard Institute for Space Studies, an atmospheric scientist who has consistently spoken out in public about the danger of climatic change. The tactics here were to work with the fact that he was a government scientist, so that almost everything he did or said could be denounced as being improper use of government time, which he should be spending tied to the workbench. He kept his job, just, but was subjected to massive amounts of harassment, including by his own employers, which was detailed in an informative and terrifying book about censorship in science.[8]

In the UK the main vehicle for the deniers has been a sinister organization set up in 2009 by Lord Lawson, a former Chancellor of the Exchequer. Called the Global Warming Policy Foundation, it refuses to reveal the sources of its funding. The director is Benny Peiser, whose previous climatic qualifications consisted of being a lecturer in Sport Science at Liverpool John Moores University. Despite its secretiveness and the lack of scientific credibility of its staff, it has achieved extraordinary success in turning the UK's present government away from its claimed intention of being 'the greenest government ever' to the point where measures against climate change are described as 'green crap'. In 2009 there was 'Climategate'. Thousands of private emails at the Climatic Research Unit at the University of East Anglia, one of the world's most respected climate research centres, were deliberately hacked and then rapidly scanned by a professional hacking organization with an office in Russia but funded from who knows where. A few emails that were mildly embarrassing were trumpeted by the unscrupulous press as if a major conspiracy had been uncovered. The real conspiracy, which was the hacking attack, went uninvestigated and unpunished.

My own experience of personal attacks started in 2012. In September 2012 the summer sea ice reached its lowest area yet, and the BBC made a film about the retreat, in which I was interviewed, among others, and satellite maps of the retreat were shown. The programme was televised on 5 September 2012, and was followed by a studio discussion in which the BBC decided that both 'sides' should be represented. The entire body of climate scientists was represented by Natalie Bennett, newly appointed chair of the Green Party, whose heart was in the right place but who had no knowledge of the Arctic. The tiny group of deniers was represented by Peter Lilley MP, formerly a Tory government minister, who had just published a report funded by the Lawson foundation in which he recommended taking no action on climate change and ignoring the Stern review. He claimed that he had been brought to the BBC under false pretences, that the BBC report had been concocted (despite the fact that satellite images of ice retreat were shown), and that I was a 'well known alarmist', a slander which he repeated five times. He claimed that he knew more about climate change than I did because he was able to quote from the 2007 IPCC assessment which concluded that summer sea ice would not disappear until the end of the twenty-first century. Despite being vice-president of an oil company, Tethys Petroleum, which works mainly with regimes in Central Asia, Lilley was subsequently put on the Environment and Climate Committee of the House of Commons, from which advantageous position he worked to define climate change legislation. Thus Lord Lawson's secretive foundation gained a representative on a powerful government committee. Lilley is not alone – there are many others in similar positions, especially in the ranks of the Republican Party in the USA, but he does personify the forces of obfuscation and misrepresentation which result in public inaction in the face of a major threat to human survival.

On the rare occasions when it engages in debate, Lawson's Global Warming Policy Foundation now adopts a position slightly modified from its earlier simple blanket denial of climate change. It agrees that the climate may be changing, though not admitting that it is due to the activities of Man, but says that the way to deal with it is adaptation, not mitigation. 'Mitigation' means trying to do something about the causes of climate change, whether by reducing emissions, trying to find ways of removing greenhouse gases from the atmosphere, or managing solar radiation by geoengineering. 'Adaptation' means, in effect, 'let's let it go and just try to live with it'. The trouble is that the amount of warming that will occur if we just let things go, which even conservative IPCC models estimate as being 4°C by the century's end, is going to be catastrophic for the maintenance of life on Earth. The warming will continue beyond 2100 and reach greater heights in the absence of action on CO_2.

Scientists who publicly state the facts about the climatic threat to the world offer a challenge to the national security of the state and provoke a response. In the UK Ian Boyd, the chief scientific adviser at the Department for Environment, Food and Rural Affairs (DEFRA), said that scientists should avoid 'suggesting that policies are either right or wrong' and should express their views 'by working with embedded advisers (such as myself), and by being the voice of reason, rather than dissent, in the public arena'. This statement of staggering arrogance assumes that Boyd's wisdom is superior to everyone else's and that he will always 'speak truth to power'. But the attitude that it represents is being enforced through directives to scientists with UK Government research contracts. Governments of countries such as Canada and Australia, until recent political changes, went even further than the UK's in their suppression of science, firing large numbers of environmental scientists so that research which determines the magnitude of climate-induced changes was simply not conducted any more. The key decisions about saving the world from climate change obviously have to be made by governments. But, tragically, some governments seem to have no intention of making them and are more interested in suppressing scientific research if results imply dissent.

The climate change deniers' emphasis on adaptation has been powerfully answered by Professor Robert P. Abele:

> As we inflict violence on the planet to the point of its mortality, we inflict violence on ourselves, to the point of our mortality. A dead planet will result in dead people, and a people and/or its leaders who are psychologically and/or ethically desensitized to the consequences of this Terran violence have no chance of long-term survival.

Or, as Chief Seattle put it more eloquently more than a century ago:

> All things are connected. Whatever befalls the Earth befalls the children of the Earth.

If we destroy our planet we destroy ourselves. There is nowhere else for us to go. There is no planet B. It will not just be farewell to ice, but farewell to life.

TIME FOR BATTLE

I assume that most readers of this book will be concerned and intelligent citizens, not necessarily scientists. What can we do, both individually and collectively, to try to save the world? There is a massive list, of course, but I will pick out a few actions that might make a real difference.

First, counter with all the power at your disposal the sewage-flow of lies and deceit emitted by climate change deniers and others who wish us to do nothing and hope that it all goes away. It will not go away. Be especially vigilant of the sinuous misrepresentations of politicians, from prime ministers downwards, and look out for glaring anomalies between what they say and what they do. When they sign up to a solemn international agreement in Paris to radically curb carbon emissions, then withdraw the feed-in tariff on solar power, fail to support renewable energy research and development, and seek to expand fossil fuel use through fracking, you know they are hypocrites and you can point out to your elected representatives that they will lose your vote unless they shape up. Scientists who study climate change should be among the first to speak up, and should be prepared to risk the blighting of their careers and the absence of establishment honours. At least they will no longer be burned at the stake, and, as the reality of climate change begins to bite, scientists who have had the courage to speak out will be respected rather than being abused and threatened.

Second, in your own life adopt every possible measure that will reduce unnecessary energy use, especially of fossil fuels. Why are more homes not insulated? This is the most energy-effective thing that you can do to your house, and from time to time a reluctant government even offers grants to assist you. Drive an economical car or ride a bike – many commutes and other types of journey in a town or city can be managed very effectively by electric bicycle. Install solar panels on your roof, even if you don't receive a feed-in subsidy.

Third, on a national scale, insist that the government changes the basis of power generation. Britain is particularly remiss in this respect. In 2015 82 per cent of our energy still came from fossil fuels.

We are world leaders in the inventive development of wave power and current turbines, and have the marine environment to exploit these new ideas, whether it be our wave-lashed west coast, the fast-flowing currents between the Orkneys, or the Severn bore. Yet only pitiful amounts of funding support comes from the government for the pioneers of these new energy systems, as I pointed out in *Underwater Technology* magazine.[9] Only recently, innovative and deserving wave power companies have closed down for lack of support.[10] The UK has huge wind resources, but has never even tried to manufacture wind turbines, leaving this to Denmark. Solar photovoltaic power is becoming cheaper all the time, and is suitable not only for home use but for larger solar farms, even in the grey UK. The problem of energy storage, which is a real one (the sun doesn't shine at night), is near solution, both from bigger batteries and from flow conversion systems, which store energy in chemical fluids contained in external tanks, which work something like fuel cells and which can store massive quantities of energy limited only by tank size. A Harvard laboratory led by Professor Michael Aziz came up with a successful flow conversion system in 2014 using quinones (organic compounds) as fluids.[11] All that is needed to put such schemes into practical operation is whole-hearted support from governments. Any plea (as in the UK) that there is no money because of austerity is bogus, because renewable energy is – in fact, has to be – the energy source of the future, so we have to adapt to it and should lead the change so that our own industry can build the new technology.

Still on a national scale, do not be afraid of nuclear power. It really is a powerful source of the baseline energy that will keep the lights on without carbon emission. Be afraid of the cack-handed British approach, which has us buying outdated and dangerous water-cooled reactors from the French (or is it the Chinese?) which will take a decade to build. All the terrible nuclear accidents that have occurred over the past forty years – Three Mile Island, Chernobyl, Fukushima – originated in the complicated cooling systems used in water-moderated reactors. Two much better pathways lie open. The *pebble bed reactor* was invented by a German consortium in the 1960s, and is basically a tower, into the top of which are placed fuel elements fused into inert pebbles. The reaction takes place within the tower, with gas cooling, and the used pebbles are dumped from the bottom. It is very simple, so unlikely to have a failure, and reactors of this kind can be built in a wide range of sizes, from giant power stations to small local energy systems. South Africa developed this system further but then stopped, but China is going ahead. The other opportunity is the *thorium reactor*, which uses thorium-232 as its fissile material. In the early days of nuclear power this was a strong competitor of uranium. Uranium reactors only became universal because the original designs were based on military submarine reactors, which have to use uranium to achieve rapid flexibility in power settings.[12] Thorium is cheaper than uranium and has the advantage that its fission products are of no military use, so there is no problem about such reactors being used by regimes whose politics worry us.

On an international scale, as I have said, the overwhelmingly important need is to undertake a colossal scientific and technical research programme on geoengineering and on carbon dioxide removal. Geoengineering is necessary to hold warming back, because we are unlikely to reduce our carbon emissions fast enough, but there are huge questions of science, engineering and governance which need to be solved before we can proceed safely. We could, of course, simply build some cloud brightening systems and/or some aerosol distribution networks and try them out. Stephen Salter, for instance, has devised a sensitive test to see whether a vapour injection system is actually having a detectable effect or not. But if we want to be safer we have to develop a research programme on modelling the impact of geoengineering techniques before we start to deploy them on a large scale.

Most important of all is the need to find a way to remove carbon dioxide from the atmosphere. This is the only thing that we can really do to save the world, so we had better do it while we still have the technical capacity and the civilization to sustain it. I have shown all the drawbacks to the various indirect techniques of CO_2 removal that have been suggested, from crushed rocks to biochar to afforestation and BECCS. The only one that can really save us is the direct removal of CO_2 from the atmosphere through some device which sucks ordinary air in at one end and emits it again at the other minus its CO_2 content, and does so at a less than impossible price. It is a problem in chemistry, physics and technology, a giant problem, but not one that is greater than that of building a huge bomb out of a reaction which previously was only observed among single atoms in a laboratory. It is the most important problem that the world faces. If we solve it, our human civilization can continue, and we can devote our energies to all our other myriad challenges, from overpopulation to water and food shortages, disease and war. If we don't solve it, we are finished. Along the way we will have said a farewell to ice, but if we stabilize our atmosphere and climate the ice may return for our descendants to wonder at and enjoy.

References

1. INTRODUCTION

1 Wadhams, P. (1990), Evidence for thinning of the Arctic ice cover north of Greenland. *Nature*, **345**, 795–7.
2 Rothrock, D. A., Y. Yu and G. A. Maykut (1999), Thinning of the Arctic sea-ice cover. *Geophysical Research Letters*, **26**, 3469–72; Wadhams, P. and N. R. Davis (2000), Further evidence of ice thinning in the Arctic Ocean. *Geophysical Research Letters*, **27**, 3973–5.
3 Wadhams, P. (2009), *The Great Ocean of Truth*. Ely: Melrose Books.
4 Headland, R. K. (2016), Transits of the Northwest Passage to end of the 2013 navigation season. Atlantic Ocean – Arctic Ocean – Pacific Ocean. *Il Polo*, **71** (3), in press.
5 Rothrock, et al., Thinning of the Arctic sea-ice cover.
6 'The ice is in a "death spiral" and may disappear in the summers within a couple of decades', M. Serreze, in *National Geographic News*, 17 Sept. 2008; 'There are claims coming from some communities that the Arctic sea ice is recovering, is getting thicker again. That's simply not the case. It's continuing down in a death spiral'. M. Serreze, Statement to *Climate Progress*, 9 Sept. 2010.

2. ICE, THE MAGIC CRYSTAL

1 Pauling, L. (1935), The structure and entropy of ice and other crystals with some randomness of atomic arrangement. *Journal of the American Chemical Society*, **57**, 2680–84.
2 Hobbs, P. V. (1974), *Ice Physics*. Oxford: Clarendon Press. See also Petrenko, V. F. and R. W. Whitworth (1999), *Physics of Ice*. Oxford: Oxford University Press; Chaplin, M. (2016), Water structure and science. www.lsbu.ac.uk/water/ice_phases.html.

3 Weeks, W. F. and S. F. Ackley (1986), The growth, structure and properties of sea ice. In Norbert Untersteiner, ed., *The Geophysics of Sea Ice*, New York: Plenum, pp. 9–164.

4 Woodworth-Lynas, C. and J. Y. Guigné (2003), Ice keel scour marks on Mars: evidence for floating and grounding ice floes in Kasei Valles. *Oceanography*, **16** (4), 90–97.

3. A BRIEF HISTORY OF ICE ON PLANET EARTH

1 Kirschvink coined the phrase Snowball Earth in a short paper, Kirschvink, J. L. (1992), Late Proterozoic low-latitude global glaciation: the snowball Earth. In J. W. Schopf and C. Klein, eds., *The Proterozoic Biosphere – a Multidisciplinary Study*. Cambridge: Cambridge University Press, pp. 51–2. Subsequent strong support for Snowball Earth came from Hoffman, P. F., A. J. Kaufman, G. P. Halverson and D. P. Schrag (1998), A Neoproterozoic snowball Earth. *Science*, **281**, 1342–6.

2 Turco, R. P., O. B. Toon, T. P. Ackerman, J. B. Pollack and Carl Sagan (1983), Nuclear Winter: Global consequences of multiple nuclear explosions. *Science*, **222** (4630), 1283–92.

4. THE MODERN CYCLE OF ICE AGES

1 Stothers, R. B. (1984), The Great Tambora eruption in 1815 and its aftermath. *Science*, **224** (4654), 1191–8.

2 Croll, J. (1875), *Climate and Time in their Geological Relations; a Theory of Secular Changes of the Earth's Climate*. Reprinted 2013 by Cambridge University Press, Cambridge Library Collection.

3 Wasdell, D. (2015), *Facing the Harsh Realities of Now*. www.apollo-gaia.org.

4 Mann, M. E., R. S. Bradley and M. K. Hughes (1999), Northern hemisphere temperatures during the past millennium: inferences, uncertainties and limitations. *Geophysical Research Letters*, **26**, 759–62.

5 Arenson, S. (1990), *The Encircled Sea. The Mediterranean Maritime Civilisation*. London: Constable.

6 Tzedakis, P. C., J. E. T. Charnell, D. A. Hodell, H. F. Kleinen and L. C. Skinner (2012), Determining the natural length of the current interglacial. *Nature Geoscience*, doi:10.1038/ngeo1358.

7   Ganopolski, A., R. Winkelmann and H. J. Schellnhuber (2016), Critical insolation – CO_2 relation for diagnosing past and future glacial inception. *Nature*, doi:10.1038/nature 16494.

5. THE GREENHOUSE EFFECT

1   Houghton, Sir John (2015), *Global Warming: The Complete Briefing*, 5th edn. Cambridge: Cambridge University Press.
2   Arrhenius, S. (1896), On the influence of carbonic acid in the air upon the temperature of the ground. *Philosophical Magazine and Journal of Science*, 41, 237–76.
3   Wasdell, D. (2014), *Sensitivity and the Carbon Budget: The Ultimate Challenge of Climate Science*. www.apollo-gaia.org.
4   Farman, J. C., B. G. Gardiner and J. D. Shanklin (1985), Large losses of total ozone in Antarctica reveal seasonal ClOx/NOx interaction. *Nature*, 315, 207–10.
5   Norval, M., R. M. Lucas, A. P. Cullen, F. R. de Grulil, J. Longstreth, Y. Takizawa and J. C. van der Leun (2011), The human health effects of ozone depletion and interactions with climate change. *Photochem. Photobiol. Sci.*, 10 (2), 199–225.
6   Molina, M. J. and F. S. Rowland (1974), Stratospheric sink for chlorofluoromethanes: chlorine atom-catalysed destruction of ozone. *Nature*, 249, 810–12. There is a more complete account in Rowland, F. S. and M. J. Molina (1975), Chlorofluoromethanes in the environment. *Reviews of Geophysics and Space Physics*, 13, 1–35.
7   Wasdell, D. (2015), *Facing the Harsh Realities of Now*. www.apollo-gaia.org.
8   Screen, J. A. and I. Simmonds (2010), The central role of diminishing sea ice in recent Arctic temperature amplification. *Nature*, 464, 1334–7.

6. SEA ICE MELTBACK BEGINS

1   Scoresby, William Jr (1820), *An Account of the Arctic Regions With a History and Description of the Greenland Whale-Fishery*. 2 vols. London: Constable (reprinted 1968, David and Charles, Newton Abbot).
2   Kelly, P. M. (1979), An Arctic sea ice data set 1901–1956. *Glaciological Data*, 5, 101–6, World Data Center for Glaciology, Boulder, Colo.
3   Parkinson, C. L., J. C. Comiso, H. J. Zwally, D. J. Cavalieri, P. Gloersen and W. J. Campbell (1987), *Arctic Sea Ice, 1973–1976: Satellite*

Passive-Microwave Observations. Washington, DC: National Aeronautics and Space Administration, SP-489.

4 Wadhams, P. (1981), Sea-ice topography of the Arctic Ocean in the region 70°W to 25°E. *Phil. Trans. Roy. Soc., London,* A302 (1464), 45–85; Comiso, J. C., P. Wadhams, W. B. Krabill, R. N. Swift, J. P. Crawford and W. B. Tucker (1991), Top/bottom multisensor remote sensing of Arctic sea ice. *Journal of Geophysical Research,* 96 (C2), 2693–709.

5 Wadhams, P. (1990), Evidence for thinning of the Arctic ice cover north of Greenland. *Nature,* 345, 795–7.

6 Rothrock, D. A., Y. Yu and G. A. Maykut (1999), Thinning of the Arctic sea-ice cover. *Geophysical Research Letters,* 26, 3469–72.

7 Wadhams, P. and N. R. Davis (2000), Further evidence of ice thinning in the Arctic Ocean. *Geophysical Research Letters,* 27, 3973–5.

8 Polyakov, I. V., J. Walsh and R. Kwok (2012), Recent changes of Arctic multiyear sea-ice coverage and the likely causes. *Bulletin of the American Meteorological Society,* doi: 10.1175/BAMS-D-11-00070.1.

9 Morello, S. (2013), Summer storms bolster Arctic ice. *Nature,* 500, 512.

10 Parkinson, C. L. and J. C. Comiso (2013), On the 2012 record low Arctic sea ice cover. Combined impact of preconditioning and an August storm. *Geophysical Research Letters,* 40, 1–6.

11 Zhang, J., R. Lindsay, A. Schweiger and M. Steele (2013), The impact of an intense summer cyclone on 2012 Arctic sea ice extent. *Geophysical Research Letters,* 40 (4), 720–26.

12 Maslowski, W., J. C. Kinney, M. Higgins and A. Roberts (2012), The future of Arctic sea ice. *Annual Reviews of Earth and Planetary Science,* 40, 625–54.

13 Macovsky, M. L. and G. Mechlin (1963), A proposed technique for obtaining directional wave spectra by an array of inverted fathometers. In *Ocean Wave Spectra,* Proceedings of a Conference held at Easton, Maryland, 1–4 May 1961. Englewood Cliffs: Prentice-Hall, pp. 235–45.

14 Wadhams, P. (1978), Wave decay in the marginal ice zone measured from a submarine. *Deep-Sea Research,* 25 (1), 23–40.

15 MIZEX Group (33 authors, inc. P. Wadhams) (1986), MIZEX East: The summer marginal ice zone program in the Fram Strait/Greenland Sea. *EOS, Transactions of the American Geophysical Union,* 67 (23), 513–17.

7. THE FUTURE OF ARCTIC SEA ICE – THE DEATH SPIRAL

1 Laxon, S. W. et al. (2013), CryoSat-2 estimates of Arctic sea ice thickness and volume. *Geophysical Research Letters*, 40, 732–7.

2 Rothrock, D. A., D. B. Percival and M. Wensnahan (2008), The decline in Arctic sea-ice thickness: separating the spatial, annual and interannual variability in a quarter century of submarine data. *Journal of Geophysical Research Oceans*, 113, C05003.

3 Kwok, R. (2009), Outflow of Arctic Ocean sea ice into the Greenland and Barents Seas: 1979–2007. *Journal of Climate*, 22, 2438–57; Polyakov, I. V., J. Walsh and R. Kwok (2012), Recent changes of Arctic multiyear sea-ice coverage and the likely causes. *Bulletin of the American Meteorological Society*, doi: 10.1175/BAMS-D-11-00070.1.

4 Tietsche, S., D. Notz, J. H. Jungclaus and J. Marotzke (2011), Recovery mechanisms of Arctic summer sea ice. *Geophysical Research Letters*, 38, L02707.

5 IPCC (2013), *Climate Change 2013. The Physical Science Basis. Working Group 1 Contribution to the Fifth Assessment Report of the Intergovernmental Panel on Climate Change. Summary for Policymakers.* Cambridge: Cambridge University Press, p. 21.

6 Wadhams, P. (2014), The 'Hudson-70' Voyage of Discovery: First Circumnavigation of the Americas. In D. N. Nettleship, D. C. Gordon, C. F. M. Lewis and M. P. Latremouille, *Voyage of Discovery, Fifty Years of Marine Research at Canada's Bedford Institute of Oceanography.* Dartmouth: BIO-Oceans Association, pp. 21–8.

7 Humpert, M. (2014), Arctic Shipping: an analysis of the 2013 Northern Sea Route season. *Arctic Yearbook 2014*, Calgary: Arctic Institute of North America. See also Arctic Council (2009), *Arctic Marine Shipping Assessment 2009 Report.*

8 National Research Council of the National Academies (2014), *Responding to Oil Spills in the U. S. Arctic Marine Environment.* Washington, DC: National Academies Press.

9 Wadhams, P. (1976), Oil and ice in the Beaufort Sea. *Polar Record*, 18 (114), 237–50.

8. THE ACCELERATING EFFECTS OF ARCTIC FEEDBACKS

1 Maykut, G. A. and N. Untersteiner (1971), Some results from a time-dependent thermodynamic model of Arctic sea ice. *Journal of Geophysical Research*, **76** (6), 1550–75.

2 Perovich, D. K. and C. Polashenski (2012), Albedo evolution of seasonal Arctic sea ice. *Geophysical Research Letters*, **39** (8), doi:10.1029/2012GL051432.

3 Pistone, K., I. Eisenman and V. Ramanathan (2014), Observational determination of albedo decrease caused by vanishing Arctic sea ice. *Proceedings of the National Academy of Sciences*, **111** (9), 3322–6.

4 Rignot, E. and P. Kanagaratnam (2006), Changes in the velocity structure of the Greenland ice sheet. *Science*, **311** (5763), 986–90.

5 McMillan, M., A. Shepherd, A. Sundal, K. Briggs, A. Muir, A. Ridout, A. Hogg and D. Wingham (2014), Increased ice losses from Antarctica detected by CryoSat-2. *Geophysical Research Letters*, **41**, 3899–905.

6 Wadhams, P. and W. Munk (2004), Ocean freshening, sea level rising, sea ice melting. *Geophysical Research Letters*, **31**, L11311, doi:101029/2004GLO20039.

7 Quinn, P. K., A. Stohl, A. Arneth, T. Berntsen, J. F. Burkhart, J. Christensen, M. Flanner, K. Kupiainen, H. Lihavainen, M. Shepherd, V. Shevchenko, H. Skov and V. Vestreng (Arctic Monitoring and Assessment Programme (AMAP)) (2011), *The Impact of Black Carbon on Arctic Climate*. Oslo: Arctic Monitoring and Assessment Programme (AMAP).

9. ARCTIC METHANE, A CATASTROPHE IN THE MAKING

1 Westbrook, G. K. et al. (2009), Escape of methane gas from the seabed along the West Spitsbergen continental margin. *Geophysical Research Letters*, **36** (15), doi: 10.1029/2009GL039191.

2 Shakhova, N., I. Semiletov, A. Salyk and V. Yusupov, (2010), Extensive methane venting to the atmosphere from sediments of the East Siberian Arctic Shelf. *Science*, **327**, 1246.

3 Dmitrenko, I. A., S. A. Kirillov, L. B. Tremblay, H. Kassens, O. A. Anisimov, S. A. Lavrov, S. O. Razumov and M. N. Grigoriev (2011), Recent changes in shelf hydrography in the Siberian Arctic: Potential

for subsea permafrost instability. *Journal of Geophysical Research*, 116, C10027, doi:10.1029/2011JC007218.

4 Shakhova, N., I. Semiletov, I. Leifer, V. Sergienko, A. Salyuk, D. Kosmach, D. Chernykh, C. Stubbs, D. Nicolsky, V. Tumskoy and Ö Gustafsson (2013), Ebullition and storm induced methane release from the East Siberian Arctic Shelf. *Nature Geoscience*, **7**, doi: 0.1038/NGEO2007; Frederick, J. M. and B. A. Buffett (2014), Taliks in relict submarine permafrost and methane hydrate deposits: Pathways for gas escape under present and future conditions. *Journal of Geophysical Research Earth Surface*, 119, 106–22, doi:10.1002/2013JF002987.

5 Whiteman, G., C. Hope and P. Wadhams (2013), Vast costs of Arctic change. *Nature*, **499**, 401–3.

6 Hope, C. (2013), Critical issues for the calculation of the social cost of CO_2: why the estimates from PAGE09 are higher than those from PAGE2002. *Climatic Change*, 117, 531–43.

7 Stern, Sir Nicholas (2006), *The Economics of Climate Change*. London: HM Treasury.

8 Overduin, P. P., S. Liebner, C. Knoblauch, F. Günther, S. Wetterich, L. Schirrmeister, H. W. Hubberten and M. N. Grigoriev (2015), Methane oxidation following submarine permafrost degradation: Measurements from a central Laptev Sea shelf borehole. *Journal of Geophysical Research. Biogeosciences*, 120, 965–78, doi:10.1002/2014JG002862.

9 Janout, M., J. Hölemann, B. Juhls, T. Krumpen, B. Rabe, D. Bauch, C. Wegner, H. Kassens and L. Timokhov (2016), Episodic warming of near bottom waters under the Arctic sea ice on the central Laptev Sea shelf. *Geophysical Research Letters*, January 2016, doi: 10.1002/2015GL066565.

10 Nicolsky, D. J., V. E. Romanovsky, N. N. Romanovskii, A. L. Kholodov, N. E. Shakhova and I. P. Semiletov (2012), Modeling sub-sea permafrost in the East Siberian Arctic shelf: The Laptev Sea region. *Journal of Geophysical Research*, 117, F03028, doi:10.1029/2012JF002358.

10. STRANGE WEATHER

1 Francis, J. A. and S. J. Vavrus (2012), Evidence linking Arctic amplification to extreme weather in mid-latitudes. *Geophysical Research Letters*, **39**, L06801, doi:.10.1029/2012GL051000.

2 Overland, J. E. (2016), A difficult Arctic science issue: mid-latitude weather linkages. *Polar Science*, in press.

3 National Academy of Sciences (2014), *Linkages Between Arctic Warming and Mid-Latitude Weather Patterns*. Washington, DC: National Academies Press.

4 Cohen, J., J. A. Screen, J. C. Furtado, M. Barlow, D. Whittleston, D. Coumou, J. Francis, K. Dethloff, D. Entekhabi, J. Overland and J. Jones (2014), Recent Arctic amplification and extreme mid-latitude weather. *Nature Geoscience*, 7 (9), 627–37, doi:10.1038/ngeo2234.

5 Ghatak, D., A. Frei, G. Gong, J. Stroeve and D. Robinson (2012), On the emergence of an Arctic amplification signal in terrestrial Arctic snow extent. *Journal of Geophysical Research*, 115, D24105.

6 Overland, J. E. and M. Wang (2010), Large-scale atmospheric circulation changes are associated with the recent loss of Arctic sea ice. *Tellus A*, 62, 1–9.

7 Liu, J., C. A. Curry, H. Wang, M. Song and R. M. Horton (2012), Impact of declining Arctic sea ice on winter snowfall. *Proceedings of the National Academy of Sciences*, 109, 4074–9, doi: 10.1073/pnas.1114910109.

8 Screen, J. A. and I. Simmonds (2013), Exploring links between Arctic amplification and mid-latitude weather. *Geophysical Research Letters*, 40, 959–64, doi: 10.1002/grl.50174.

9 Grassi, B., G. Redaelli and G. Visconti (2013), Arctic sea-ice reduction and extreme climate events over the Mediterranean region. *Journal of Climate*, 26, 10101–10, doi:10.1175/JCLI-D-12-00697.1.

10 Wu, B., D. Handorf, K. Dethloff, A. Rinke and A. Hu (2013), Winter weather patterns over northern Eurasia and Arctic sea ice loss. *Monthly Weather Review*, 141, 3786–800, doi:10.1175/MWR-D-13-00046.1.

11 Wilkins, Sir Hubert (1928), *Flying the Arctic*. New York: Grosset and Dunlap.

12 Haberl, H., D. Sprinz, M. Bonazountas, P. Cocco, Y. Desaubies, M. Henze, O. Hertel, R. K. Johnson, U. Kastrup, P. Laconte, E. Lange, P. Novak, I. Paavolam, A. Reenberg, S. van den Hove, T. Vermeire, P. Wadhams and T. Searchinger (2012), Correcting a fundamental error in greenhouse gas accounting related to bioenergy. *Energy Policy*, 45, 18–23.

13 Arnell, N. W. and B. Lloyd-Hughes (2014), The global-scale impacts of climate change on water resources and flooding under new climate and socio-economic scenarios. *Climatic Change*, 122, 1–2, 127–40, doi: 10.1007/s10584-013-0948-4.

11. THE SECRET LIFE OF CHIMNEYS

1 Marshall, J. and F. Schott (1999), Open-ocean convection: observations, theory and ideas. *Reviews of Geophysics*, 37, 1–63.

2 Scoresby, William Jr (1820), *An Account of the Arctic Regions With a History and Description of the Greenland Whale-Fishery*. 2 vols. London: Constable (reprinted 1968, David and Charles, Newton Abbot).

3 Wilkinson, J. P. and P. Wadhams (2003), A salt flux model for salinity change through ice production in the Greenland Sea, and its relationship to winter convection. *Journal of Geophysical Research*, 108 (C5), 3147, doi:10.1029/2001JC001099.

4 MEDOC Group (1970), Observations of formation of deep-water in the Mediterranean Sea, 1969. *Nature*, 227, 1037–40.

5 Wadhams, P., J. Holfort, E. Hansen and J. P. Wilkinson (2002), A deep convective chimney in the winter Greenland Sea. *Geophysical Research Letters*, 29 (10), doi:10.1029/2001GL014306.

6 Budéus, G., B. Cisewski, S. Ronski, D. Dietrich and M. Weitere (2004), Structure and effects of a long lived vortex in the Greenland Sea. *Geophysical Research Letters*, 31, L053404, doi:10.1029/2003 62 017983.

7 Wadhams, P., G. Budéus, J. P. Wilkinson, T. Loyning and V. Pavlov (2004), The multi-year development of long-lived convective chimneys in the Greenland Sea. *Geophysical Research Letters*, 31, L06306, doi:10.1029/2003GL019017.

8 Wadhams, P. (2004), Convective chimneys in the Greenland Sea: a review of recent observations. *Oceanography and Marine Biology. An Annual Review*, 42, 1–28.

9 De Jong, M. F., H. M. Van Aken, K. Våge and R. S. Pickart (2012), Convective mixing in the central Irminger Sea: 2002–2010. *Deep-Sea Research, I*, 63, 36–51.

12. WHAT'S HAPPENING TO THE ANTARCTIC?

1 See website climate.nasa.gov/news/.

2 Rignot, E., J. L. Bamber, M. R. van den Broeke, C. Davis, Y. Li, W. J. van de Berg and E. van Meijgaard (2008), Recent Antarctic ice mass loss from radar interferometry and regional climate modelling. *Nature Geoscience*, 1 (2), 106–10.

3 Wadhams, P., M. A. Lange and S. F. Ackley (1987), The ice thickness distribution across the Atlantic sector of the Antarctic Ocean in

midwinter. *Journal of Geophysical Research*, 92 (C13), 14535–52; Lange, M. A., S. F. Ackley, P. Wadhams, G. S. Dieckmann and H. Eicken (1989), Development of sea ice in the Weddell Sea Antarctica. *Annals of Glaciology*, 12, 92–6.

4 Wadhams et al. (1987), The ice thickness distribution across the Atlantic sector of the Antarctic Ocean in midwinter.

5 Ibid.

6 Wadhams, P. and D. R. Crane (1991), SPRI participation in the Winter Weddell Gyre Study 1989. *Polar Record*, 27 (160), 29–38.

7 Ackley, S. F., V. I. Lytle, B. Elder and D. Bell (1992), Sea-ice investigations on Ice Station Weddell. 1: ice dynamics. *Antarctic Journal of the US*, 27, 111–13.

8 Hellmer, H. H., M. Schröder, C. Haas, G. S. Dieckmann and M. Spindler (2008), Ice Station Polarstern (ISPOL). *Deep-Sea Research II*, 55, 8–9.

9 Massom, R. A., H. Eicken, C. Haas, M. O. Jeffries, M. R. Drinkwater, M. Sturm, A. P. Worby, X. Wu, V. I. Lytle, S. Ushio, K. Morris, P. A. Reid, S. G. Warren and I. Allison (2001), Snow on Antarctic sea ice. *Reviews of Geophysics*, 39, 413–45; Eicken, H., M. A. Lange, H.-W. Hubberten and P. Wadhams (1994), Characteristics and distribution patterns of snow and meteoric ice in the Weddell Sea and their contribution to the mass balance of sea ice. *Annals of Geophysics*, 12, 80–93.

10 Parkinson, C. L. and D. J. Cavalieri (2012), Antarctic sea ice variability and trends, 1979–2010, *The Cryosphere*, 6, 871–80, doi:10.5194/tc-6-871-2012.

11 Zwally, H. J., J. C. Comiso, C. L. Parkinson, W. J. Campbell, F. D. Carsey and P. Gloersen (1983), *Antarctic Sea Ice 1973–1976: Satellite Passive-Microwave Observations*. Washington, DC: NASA, Rept. SP-459.

12 Bromwich, D. H., J. P. Nicolas, A. J. Monaghan, M. A. Lazzara, L. M. Keller, G. A. Weidne and A. B. Wilson (2013), Central West Antarctica among the most rapidly warming regions on Earth. Southern ocean winter mixed layer. *Nature Geoscience*, 6, 139–45.

13 Steig, E. J., D. P. Schneider, S. D. Rutherford, M. E. Mann, J. C. Comiso and D. T. Shindell (2009), Warming of the Antarctic ice-sheet surface since the 1957 International Geophysical Year. *Nature*, 457, 459–62.

14 Bromwich et al. (2013), Central West Antarctica among the most rapidly warming regions on Earth.

15 Maksym, T., S. E. Stammerjohn, S. Ackley and R. Massom (2012), Antarctic sea ice – a polar opposite? *Oceanography*, 25, 140–51.

16 Zwally, H. J. and P. Gloersen (1977), Passive microwave images of the polar regions and research applications. *Polar Record*, 18, 431–50; Steig et al. (2009), Warming of the Antarctic ice-sheet surface.

17 Bagriantsev, N. V., A. L. Gordon and B. A. Huber (1989), Weddell Gyre – temperature maximum stratum. *Journal of Geophysical Research*, **94**, 8331–4; Gordon, A. L. and B. A. Huber (1990), Southern ocean winter mixed layer. *Journal of Geophysical Research*, **95**, 11655–72.

18 www.climate.nasa.gov/news/.

19 Zhang, J. (2014), Modeling the impact of wind intensification on Antarctic sea ice volume. *Journal of Climate*, **27**, 202–14.

20 Jacobs, S., A. Jenkins, H. Hellmer, C. Giulivi, F. Nitsche, B. Huber and R. Guerrero (2012), The Amundsen Sea and the Antarctic ice sheet. *Oceanography*, **25**, 154–63.

21 Mengel, M. and A. Levemann (2014), Ice plug prevents irreversible discharge from East Antarctica. *Nature Climate Change*, **4**, 451– 5, doi:10.1038.

22 Peterson, R. G. and W. B. White (1998), Slow oceanic teleconnections linking the Antarctic Circumpolar Wave with the tropical El Niño–Southern Oscillation. *Journal of Geophysical Research*, **103**, 24573–83.

23 Comiso J. C., R. Kwok, S. Martin and A. L. Gordon (2011), Variability and trends in sea ice extent and ice production in the Ross Sea. *Journal of Geophysical Research*, **116**, C04021, doi:10.1029/2010JC006391.

24 Rind, D., M. Chandler, J. Lerner, D. G. Martinson and X. Yuan (2001), Climate response to basin-specific changes in latitudinal temperature gradients and implications for sea ice variability. *Journal of Geophysical Research*, **106**, 20161–73.

25 Wilson, A. B., D. H. Bromwich, K. M. Hines and S.-H. Wang (2014), El Niño flavors and their simulated impacts on atmospheric circulation in the high-southern latitudes. *Journal of Climate*, **27**, 8934–55, doi:10.1175/JCLI-D-14-00296.1.

26 Francis, J. A. and S. J. Vavrus (2012), Evidence linking Arctic amplification to extreme weather in mid-latitudes. *Geophysical Research Letters*, **39**, L06801, doi:.10.1029/2012GL051000.

27 Whiteman, G., C. Hope and P. Wadhams (2013), Vast costs of Arctic change. *Nature*, **499**, 401–3.

13. THE STATE OF THE PLANET

1 Ehrlich, P. R. and A. H. Ehrlich (2014), Collapse: what's happening to our chances? http://mahb.stanford.edu/blog/collapse-whats-happening -to-our-chances?

2 UN (2015), *World Population Prospects, the 2015 Revision*. New York: United Nations Population Division, Department of Economic and Social Affairs.

3 Meadows, D. H, D. L. Meadows, J. Randers and W. W. Behrens III (1972), *The Limits to Growth*. Universe Books.

4 MacKay, Sir David J. C. (2009), *Sustainable Energy – Without the Hot Air*. UIT Cambridge Ltd. Available for download, www.without-hotair.com.

5 Paterson, Owen. The State of Nature: Environment Question Time. Conservative Party fringe, Manchester, 29 September 2013.

6 Royal Society (2009), *Geoengineering the Climate: Science, Governance and Uncertainty*. London: Royal Society.

7 Latham, J. (1990), Control of global warming? *Nature*, **347**, 339–40.

8 Salter, S., G. Sortino and J. Latham (2008), Sea-going hardware for the cloud albedo method of reversing global warming. *Philosophical Transactions of the Royal Society*, **A366**, 3989–4006.

9 Latham, J., A. Gadian, J. Fournier, B. Parkes, P. Wadhams and J. Chen (2014), Marine cloud brightening: regional applications. *Philosophical Transactions of the Royal Society*, **A372**, 20140053.

10 Rasch, P., J. Latham and C-C. Chen (2009), Geoengineering by cloud seeding: influence on sea ice and climate system. *Environmental Research Letters*, **4**, 045112, doi:10.1088/1748-9326/4/4/045112.

11 Rignot, E., J. Mouginot, M. Morlinghem, H. Senussi and B. Scheuchi (2014), Widespread, rapid grounding line retreat of Pine Island, Thwaites, Smith and Kohler Glaciers, West Antarctica, from 1992 to 2011. *Geophysical Research Letters*, **41**, 3502–9, doi:10.1002/2014GL060140.

12 Jackson, L. S., J. A. Crook, A. Jarvis, D. Leedal, A. Ridgwell, N. Vaughan and P. M. Forster (2014), Assessing the controllability of Arctic sea ice extent by sulphate aerosol geoengineering. *Geophysical Research Letters*, **42**, 1223–31, doi:10.1002/2014GL062240.

13 Crutzen, P. J. (2006), Albedo enhancement by stratospheric sulfur injections: a contribution to resolve a policy dilemma? *Climatic Change*, **77**, 211–20.

14 Xia, L., A. Robock, S. Tilmes and R. R. Neely III (2016), Stratospheric sulfate engineering could enhance the terrestrial photosynthesis rate. *Atmospheric Chemistry and Physics*, **16**, 1479–89.

15 Williamson, P. (2016), Emissions reduction: scrutinize CO_2 removal methods. *Nature*, **530**, 153–5.

16 Halter, R. (2011), *The Insatiable Bark Beetle*. Victoria BC: Rocky Mountain Books.

14. A CALL TO ARMS

1 Wasdell, D. (2015), *Facing the Harsh Realities of Now*. www.apollo-gaia.org.

2 Emmerson, C. and G. Lahn (2012), *Arctic Opening: Opportunity and Risk in the High North*. London: Chatham House/Lloyd's Risk Report. www.chathamhouse.org/sites/default/files/public/Research/Energy,%20Environment%20and%20Development/0412arctic.pdf.

3 International Monetary Fund (2013), *World Economic Outlook, April 2013*. New York: IMF.

4 Ibid.

5 Full text of speech available on website of Margaret Thatcher Foundation, www.margaretthatcher.org.

6 Houghton, J. T., G. J. Jenkins and J. J. Ephraums (eds.) (1990), *Climate Change. The IPCC Scientific Assessment*. Cambridge: Cambridge University Press.

7 Oreskes, N. and E. M. Conway (2010), *Merchants of Doubt: How a Handful of Scientists Obscured the Truth on Issues from Tobacco Smoke to Global Warming*. London: Bloomsbury Press.

8 Bowen, M. (2008), *Censoring Science: Inside the Political Attack on Dr. James Hansen and the Truth of Global Warming*. New York: Dutton Books.

9 Wadhams, P. (2015), New roles for underwater technology in the fight against catastrophic climate change. *Underwater Technology*, 33 (1), 1–2.

10 Merry, S. (2016), Outlook for the wave and tidal stream industry in the UK. *Underwater Technology*, 33 (3), 139–40.

11 Huskinson, B., M. P. Marshak, C. Suh, E. Süleyman, M. R. Gerhardt, C. J. Galvin, X. Chn, A. Asparu-Guzik, R. G. Gordon and M. J. Aziz (2014), A metal-free organic–inorganic aqueous flow battery. *Nature*, 505, 195–8; Lin, K. et al. (2015), Alkaline quinone flow battery. *Science*, 349, 1529.

12 Martin, R. (2012), *Superfuel. Thorium, the Green Energy Source for the Future*. London: Palgrave Macmillan.

Index

Abele, Professor Robert P., 202
ADCP (acoustic döppler current
  profiler), 150
aerosols, 58, 59, 61, 178, 181–2, 205
afforestation, 183, 184, 185, 205
Africa, 46, 127, 134, 140–41, 142,
  172–3, 180, 182
Agassiz, Lake, 41
agriculture, 41, 42–3, 45, 56–7, 196
  artificial fertilizers, 57, 172
  and BECCS concept, 184
  and extreme weather events, 134,
    139–42, 172–3
  global food production, 105, 134,
    140–42, 172–3, 176–7, 184, 196
Alaska, 2, 71, 93, 110
albedo (reflectivity)
  and Antarctic, 156, 169
  and black carbon feedback, 116, 118
  and climatic feedbacks, 105–9, 115,
    118, 156, 170
  of Earth as a whole, 25, 47–8
  and Earth's balance of radiation, 58
  and forests, 185
  negative feedback, 39
  of open water, 4, 105, 106–7
  and sea ice, 4, 15, 25, 105–7, 118
  of snow, 15, 25, 76–7, 105, 106
  during 'Snowball Earth', 24, 25
  snowline retreat feedback, 106,
    107–9, 108fig, 115, 118, 156, 170

Alfred Wegener Institute,
  Bremerhaven, 150, 152, 159
Amundsen, Roald, 2, 92–3
Amundsen Sea, 163, 165
Andersonian College and Museum,
  Glasgow, 34
Antarctic Ocean see Southern Ocean
Antarctic Treaty (1959), 159
Antarctica
  and albedo feedback, 156, 169–70
  Antarctic Peninsula, 2, 113, 157,
    163, 165
  collapse of Larsen B ice shelf, 157
  and global weather patterns, 157
  and Ice Ages, 30, 31, 46
  ice cores, 32, 34, 41
  ice sheet, 21, 32, 34, 110, 112–13,
    157, 161, 168
  importance of, 156
  and MCB method, 180
  'ozone hole' over, 57–8
  in Pliocene period, 31
  retreating ice sheet, 110, 112–13,
    157, 168
  snowfall in, 161, 162fig
  see also Southern Ocean
APLIS (Applied Physics Laboratory
  Ice Station), 72, 74, 75
Apollo-8 mission, 2
Arab Spring (2011), 141
Archimedes' Principle, 114